Forensics, Fossils and Fruitbats

For my parents
who encouraged reading, my love of science and the need to research

Forensics, Fossils and Fruitbats

A FIELD GUIDE TO AUSTRALIAN SCIENTISTS

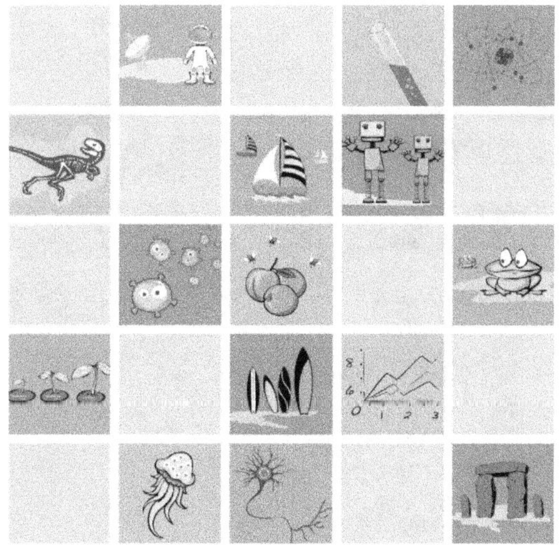

Stephen Luntz

National Library of Australia Cataloguing-in-Publication entry

Luntz, Stephen.

Forensics, fossils and fruitbats : a field guide to Australian scientists/by Stephen Luntz.

9780643097452 (pbk.)
9780643097469 (pdf)

Scientists – Australia – Biography.
Science.

509.22

Published by

CSIRO PUBLISHING
36 Gardiner Road, Clayton VIC 3168
Private Bag 10, Clayton South VIC 3169
Australia

Telephone: [+613] 9545 8555
Local call: 1300 788 000 (Australia only)
Fax: +61 3 9662 7555
Email: csiropublishing@csiro.au
Web site: www.publishing.csiro.au

Set in 10.5/13 Adobe Minion Pro and Optima

Edited by Elaine Cochrane
Cover design by Alicia Freile, Tango Media
Illustrations by Naomi Dowsett
Text design by James Kelly
Typeset by Desktop Concepts Pty Ltd, Melbourne
Printed by Ingram Lightning Source

The profiles herein are edited versions of articles previously published in *Australasian Science*.

CSIRO PUBLISHING publishes and distributes scientific, technical and health science books and journals from Australia to a worldwide audience and conducts these activities autonomously from the research activities of the Commonwealth Scientific and Industrial Research Organisation (CSIRO).

Mar26_RP_ILS

CONTENTS

Acknowledgements ix

Introduction xi

ARCHAEOLOGISTS AND PALAEONTOLOGISTS 1

Ancient fossils and future minds 2

Young scientist in an old field 5

Forensic archaeology: 'you have to be a bit weird' 9

The iceman cometh 11

A Long history 14

ASTRONOMERS AND SPACE SCIENTISTS 17

Space suits that fit 18

Asteroid wrapped 20

Engineering astronomy's future 22

A cool entry 25

A vision to Mars 28

Behind the Moon landings 31

BOTANISTS AND AGRICULTURAL SCIENTISTS 33

Top of the tree 34

A burning issue 37

Science brings bread and peace 40

CHEMISTS 43

Chemistry that's better than nature 44

On the money 46

Drug test leads to explosives 49

COMPUTER SCIENTISTS 51

Stone circles to computer scams and clinical notes 52

Computers get the joke 55

A novel scientist 57

EARTH SCIENTISTS 59

An explosion of science: I blast 'em, you mine 'em 60
Touchdown on a cold planet 63
Smoother sailing 65
Geoscientist shapes the world 67
Both sides now 70
The weather forecast is cool 73

ENGINEERS 75

Seeing hope 76
Robots ride high 79
Clean water's 'no-tech' solution 81
Dancing with the flow 84
Earth houses that don't shake down 86
The science of swimming 89

GENETICISTS 91

Forensic consulting 92
Swat the difference 94
Thoroughbred geneticist 96

MARINE BIOLOGISTS 99

Sometimes cold, always beautiful 100
Reef restoration 103
Eight eyes, no brain 105

MATHEMATICIANS 107

Mathematical art 108
The beer scheduler 110
Statistics brought to life 113

MEDICAL SCIENTISTS 115

A model scientist 116
Skin for life 118
A parasitologist and clergyman with 'a passion for poo' 121
At the viral frontline 124
Flu busting cold case 127

NEUROSCIENTISTS **131**
The brain and the Bomb 132
The brain collector 135
Music to deaf ears 137
Music and the mind 140

PHYSICISTS **143**
The stunt physicist 144
The man with X-ray vision 146
Superconducting physicist 149
A quantum of music 152
Physics made fun 155
A quantum leap for children's fiction 158
Tennis anyone? 160
Are nanoparticles safe? 162

SCIENCE COMMUNICATORS **165**
The surfing scientist 166
Taking science to the media 169
Fires, killer whales and megafauna 172
Science-ology 174

ZOOLOGISTS **177**
Do kangaroos have friends? 178
Refugee solves Australian problems 181
The real Batman 184
Animal intelligence 187
The talking ape 190
Life in the canopy 193
Marsupial nutrition 195
Kangaroos, frogs, crocodiles and rockets 198
There's a moth in my chocolate 201
Birdcatching on the fly 203
Crocs and bum-breathing turtles 205

ACKNOWLEDGEMENTS

Before all else, I would like to thank my editors. Guy Nolch, editor of *Australasian Science*, came up with the idea of the 'Cool Scientist' column, which eventually became this book. He suggested many of the scientists included here, as well as shaping my writing style. Moreover, it was Guy who gave me my start in science writing, hiring me to work at *Australasian Science* in the first place, a decision for which I am forever grateful. CSIRO Publishing showed faith in the viability of the project and Briana Melideo, John Manger, Tracey Millen and Elaine Cochrane have corrected grammar (and occasionally facts), provided ideas, and assisted with the adjustments required for transformation from a string of columns to a book. Any errors are mine, not theirs or, of course, the scientists'.

Greg Barber, Virginia Graham, Sam McGrath, Colin Smith, Kate Nairn, Michele Fountain, Emma Burrows and Sharlot Clark recommended appropriate scientists or assisted with tracking down hard-to-find people. Suggestions from Greg and Virginia helped inspire the creation of the column and subsequent contributions from the others have made my job easier and this book more interesting and diverse. My apologies to anyone I have forgotten.

I would also like to thank those who encouraged me in this process, particularly when the prospects of publication looked bleak: my parents, Richard McConachy, Heather Corinna, Emma Rush, Leah Bryant, Monica Dux and Jon Symons.

Morgan Hunter, Julia Keating, Jennifer Tarr, Anne Hunter, Linden Salter Duke and Maia Sauren provided valuable assistance during the process of transforming the columns to a book, and I am grateful for their time and advice. Too many people to list attempted to resolve the tricky question of a title.

My good fortune in marvellous housemates Cailin Howarth and Heather Madsen makes working from home a pleasure not a strain, something invaluable to any writer.

I'd particularly like to thank my business partners at Above Quota Elections; Jaimie Adam, Charles Richardson and Charlie Sanders, along with our wonderful staff, notably my most recent deputies Julia Perdevich and Haydn Steel. Without all of you I would long ago have had to find other employment that would probably have made science writing impossible.

Finally, and of course most crucially, I would like to thank the scientists and engineers who appear within these pages. Most of them are under considerable pressure to squeeze too much into the day, yet gave generously of their time for the original interviews. Most were also remarkably prompt in responding when I contacted them to update their profiles for this book. I am intensely grateful for this, and hope they feel I have done justice to them, their stories, and their science.

INTRODUCTION

I am a fierce believer in science. Science isn't perfect, but rather like democracy, the cure for its problems is usually more of it, not less.

That's not really an unusual statement. Science has been so successful at improving our lives that even its enemies, of which there are quite a few, usually feel the need to praise it. Indeed they often claim to be its greatest defenders while they're stabbing it in the back.

What's more unusual is that I'm a great believer in scientists. As individuals scientists are much further from perfect than the trade they practise, but they tend to be remarkable, fascinating, generous and multi-talented people. This is a much rarer statement. In his first set of science comedy shows, comedian Ben McKenzie (p. 174) asked his audience to name some scientists. Once they'd finished with Einstein, Darwin and Galileo he asked for *living* scientists. The list usually came to one: Steven Hawking. If you're not dying of a terrible disease you're apparently not interesting enough to be remembered.

For a long time fictitious scientists were even worse – evil geniuses trying to take over or destroy the world, leavened occasionally by figures like the Doc from *Back to the Future* – nice, but crazy old men with mad hair. It's a little better now. Although programs like *Big Bang Theory* still present a disturbing stereotype of science nerds, prime-time TV is flooded with crime-fighting forensic scientists and even an attractive mathematician pair. These shows present an utterly unrealistic vision of the resources and opportunities available to the average scientist, but I suspect viewers are even more suspicious of the characters. Given the stubborn perception of scientific personalities, *NCIS*'s Abby is seen as improbable as the show's plots.

In 2002, when I first told people I was writing a column called 'Cool Scientist' for *Australasian Science* magazine, the most common responses were 'have you found any?' or suggestions I'd run out at three. As you can see, this book contains 73, and there are quite a few banked up for interviews. And that's just Australians.

There may be no caffeine-addicted goth geniuses among them, but some are pretty cool nevertheless. Don't believe me? Ask the 120 000 women who voted Valerio Vittone (p. 116) Australia's most eligible bachelor. Or check out Saul Griffith (p. 76), who doesn't let inventing technology capable of transforming the lives of millions stop him going kite surfing. On kites he designs himself. Or Kirsty Spalding (p. 132), who actually is solving crimes when she's not unravelling the secrets of how nerve cells grow, with possible pointers to restoring brains damaged by injury or disease.

A lot of the scientists in this book were surprised anyone thought they deserved a profile. Not all of them seem comfortable in the spotlight, but the work they do can make up for that. Some are saving lives or finding ways more people can live in health and comfort. Others are taking on the great environmental challenges of our time.

For most scientists in this book the work is more basic, further from obvious applications, but history teaches us that in the long run the best basic science has a way of proving surprisingly useful. In the meantime however, every person in this book is making the world a more interesting place for all of us to live in and learn about.

Of course not all scientists are doing work as interesting as those featured in these pages. Those included here were chosen partly to represent a range of fields and stages in a scientific career. However, selection also required them to either be doing particularly exciting research, or have some other feature that made them stand out from the crowd. On the other hand, the scientists in this book are far from unique. There are dozens of other scientists I'm hoping to interview who're doing just as interesting things. I've no doubt there are hundreds, maybe thousands, more who deserve to be written up in pages such as these.

Does it matter what sort of people scientists are? I think it does. Over the last two decades the proportion of students studying science in the later years of high school has plummeted, and applications for science places at university are also well down. It's hard to know how much of this is driven by the attraction of much better pay in some other fields, and how much is an image problem. I'm fairly sure that both contribute. Bizarrely, some commentators assured us that the flight of mathematically talented students out of science and into the financial sector didn't matter, because they'd end up helping the economy anyway. Leaving aside whether the economy is the most important thing, it's now clear that all we got out of the concentration of brainpower in finance were new and better ways to create a global financial crisis.

Meanwhile, we're now seeing how hard it is to turn low science enrolments around. Fewer people with science qualifications are going into teaching, to the point where many students are taught physics at high school by teachers who never touched the topic at university. There are exceptions, but most teachers without much background in a subject struggle to enthuse their students – not surprisingly, since they'd usually rather be teaching something else. Uninspired students don't go on and do the same subjects at university, and a vicious circle is created.

Scientists complain about low pay, uncertain research funding and lack of recognition of their achievements. Most of the time they're right to do so, particularly in Australia, which has neglected science in recent years to a shocking extent. Even so, there are lots of very exciting careers in science, roles that make a real difference to the world. Properly explained, science can be a lot more thrilling than working for a bank, and rewarding in far more ways. Most of the scientists in this book love what they do, partly because it's so important. The challenges the world will face over coming decades, problems like global warming, food insecurity, aging populations and loss of biodiversity, will need more and better science to solve them. We need scientists like those in this book to take these challenges on. I hope that some readers will one day be among them.

I recommend everyone reading this book catch one of Ben's shows. I hope if he asks the question again readers can name so many living scientists he has to cut them off so he can get back to the jokes.

Archaeologists and palaeontologists

Ancient fossils and future minds

Patricia Vickers-Rich complains that there are 'only so many hours in the day,' but she still manages to pack a superhuman number of activities into her 24. Besides her outstanding career in palaeontology, Professor Vickers-Rich is director of the Monash Science Centre (MSC), a groundbreaking institution in science education. She's also making a difference for children in refugee camps, and has organised scientifically themed stamp issues for Australia Post. While fact-checking this profile she warns she'll be hard to contact for two months while off 'filming with David Attenborough and taking part in an expedition to the Lesser Himalayas'. However, she says the biggest draw on her time is now fundraising, either for field trips or for the MSC.

Vickers-Rich is famous for her work on southern Australian dinosaurs. During the Cretaceous period, about 106 to 115 million years ago, southern Australia lay well south of the Antarctic Circle. The climate was much warmer than modern Antarctica, but living things had to deal with winters completely without sunlight. The unique creatures that adapted to these conditions were almost unknown until Vickers-Rich, her husband Dr Tom Rich, and their co-workers discovered fossils along the southern coast of Victoria, from Inverloch to west of Point Otway. Their discovery was so significant it made the cover of *Time*.

Vickers-Rich and Rich discovered several new species of dinosaur, and named two after their children and one, Qantassaurus, after a sponsor. Researchers and volunteers from their laboratories at Monash University and Museum Victoria continue to dig near Inverloch where some of the discoveries were made, but Vickers-Rich no longer has much time to take part, her role being mainly fundraising and publishing.

Vickers-Rich's research focus is now on some of the oldest animal fossils from sites in Namibia and Saudi Arabia. She says that this was the place where the 'last stand' of some of the first experiments in animal life, the ediacarans, took place. She expects to continue to search for reasons why they first appeared and why they were crowded out by more advanced forms like trilobites about 542 million years ago.

She chairs the Australian arm of the International Geological Correlation Program (UNESCO), a body that helps coordinate geological and palaeontological research, and has co-led one program – IGCP493 (www.geosci.monash.edu.au/precsite) – on the ediacaran era. The book she co-authored on this time in Earth's history, *The Rise of Animals: Evolution and Diversification of the Kingdom Animalia*, won the Victorian Premier's Award for Science Writing for 2008–2009.

To Vickers-Rich, an interest in science goes back as far as she can recall. 'I don't remember my first doll, but I do remember my first grasshopper in a jar,' she says. She grew up in America, and her parents encouraged her love of learning to the point that they moved house so the family could survive when she was accepted into the University of California at Berkeley – Vickers-Rich and her parents all had to work in order to make ends meet, but they were willing to do this just to support her curiosity.

Her PhD at Columbia was on ancient Australian birds, and a Fulbright Scholarship brought her to this continent. Her PhD thesis, combined with work by Peter Murray, has since been turned into the book *Magnificent Mihirungs: The Colossal Flightless Birds of the Australian Dreamtime*, published by Indiana University Press. These birds, the dromornithids, were once the continent's dominant herbivores, surviving until about 50 000 years ago, and were more closely related to geese and swans than to emus and cassowaries.

Communicating science is a passion for Vickers-Rich, but the 14 books she has published on palaeontology are just the beginning. The MSC hosts 20 000–50 000 students a year. It connects students directly with the work of Monash and other researchers as well as hosting such crowd-pleasing exhibition topics as 'Extreme danger – the science behind natural disasters'.

The MSC also prepares science kits for primary and lower-secondary school teachers and their students, for example a kit to accompany the stamps Vickers-Rich organised, and it has provided professional development for 300 teachers. Vickers-Rich says that these teachers 'will go back to the schools and pass on their knowledge to other teachers, so it multiplies out to many students'.

One of the most remarkable features of the MSC is its work promoting science in developing countries. Vickers-Rich organised the distribution of *Animals of the World* colouring books for children in Afghanistan and Timor. These books are used in both schools and refugee camps, many of which are stretched simply ensuring their occupants are safe and fed. Attending to education or emotional needs can be beyond the camps' capacity, so Vickers-Rich tries to pack as many functions as possible into a single book. The books introduce animal species and geography, and are printed in local languages (Dari, Pasto, Portuguese, Tetun, Indonesian, Malay; even Russian and Slovenian versions have since been produced), and English. For many recipients this will be the only book they own, and the teaching kits often come with a set of pencils, a ruler and protractor supplied by the National Geographic Society, among other school tools. The care with which the children treasure these gifts and their joy in them is beyond description.

The MSC also has a program supporting the development of a national science museum in Timor-Leste. With Nobel Prize winner President Jose Ramos Horta, Vickers-Rich has co-written a children's book on the geology of the island to fire interest among Timorese children in learning about their land and its past.

Vickers-Rich explains that the MSC's developmental work coincides with research work or teaching done by Monash staff and other research scientists. 'When I was in South Africa, where Monash has a campus, I met with teachers from the townships and talked to them about how we could interact to add more to the curriculum. I have visited two schools in the last couple of years and now have a better idea what resources they have and the environment in which they are situated so we can adapt our kits.'

Vickers-Rich notes the MSC is almost unique. 'There are many science centres around the world doing fantastic work, but only the Lawrence Hall of Science at UC Berkeley is quite like us. We develop kits in concert with people from Monash who are experts as well as the teachers who will use them.'

The MSC uses staff and some volunteers effectively, enabling it to produce exhibitions far more economically than many other centres. They also recycle old exhibition materials, revamp them and use them again and again – in particular the furniture that has formed the base of a great variety of exhibitions over the years. 'We're small, and we wish to stay small because it gives us flexibility and an environment that allows individual staff members to be nurtured and be very much a part of the operation,' Vickers-Rich says. Nevertheless, the MSC's lack of funding is clearly a major concern for her.

One-quarter of the MSC's funds come from Monash University, with the rest from a mix of private sponsorship, charitable trusts, government grants and personal donations. Someone 'down the road' recently gave $10 000 to pay for the architectural plans for a desperately needed second building.

It is easier to gain grants for individual projects than recurrent funding, and Vickers-Rich dreams of a time when an endowment of $500 000 per annum could underwrite the core staff and keep the MSC functioning without having the ongoing pressure of raising that sort of funding year upon year. 'It is not an environment where one has job security and where I can offer my staff a career path. Staff have stayed with me for years, but every now and then a treasured and productive team member must look for another job where there is security, and at that point Monash loses such brilliant assets' Vickers-Rich says with some regret. Nevertheless, she refuses to let the uncertainty get in the way of a stream of new exhibitions, supervision of students, continuing her research priorities and publication of her own papers on the earliest forms of animal life. Meanwhile, she is on the hunt for that funding to ensure that the Monash Science Centre does not follow in the steps of her ediacarans – and become history.

Young scientist in an old field

Scott Hocknull is a busy man. It took two years to land an interview, but that's the price of being in an area of science that, for all its glamour, takes you into some of the most remote parts of the country. It's also the price of being Young Australian of the Year for 2002, which made Hocknull an ambassador for science as well as youth, and very much in demand.

Hocknull's first memories of his interest in palaeontology go back to when his grandmother gave him dinosaur toys at the age of six or seven.

'They weren't to scale and were flashy colours – more like dragons really – but I liked playing with them. Then when we went to the UK for a holiday, all I wanted to do was go to the British Museum of Natural History because David Attenborough was my hero and I thought he really worked there. There was a 30-metre long Diplodocus skeleton and I thought: "That's what I want to do".'

At 12 Hocknull's family moved from Darwin to near Brisbane, and regular visits to a museum became possible. Rather to the staff's surprise, Hocknull asked if he could volunteer. 'One of the great things about the Queensland Museum is the strong volunteer ethos they have,' Hocknull says. 'I was allowed to prosper because I was with a bunch of people who liked the same things I liked.'

Nevertheless, Hocknull said that the curator of vertebrate fossils at the time, Dr Ralph Molnar, 'gave me a large wake-up call about the job prospects in palaeontology, telling me I'd have to either wait for someone to die or retire, or create my own job'.

Hocknull's parents were concerned at this news, but Hocknull discovered that Molnar was 55 and asked his father what retirement age was. Nine years later, when Molnar retired, Hocknull had established such a reputation in Queensland palaeontology that he become the youngest museum curator in any Australian museum, having just completed his Honours degree.

At age 16 Hocknull became the youngest scientific author in the country when he discovered and described a species of ancient fossil freshwater bivalve. He also took part in numerous expeditions, at first by 'saving my pennies' so he could pay for himself to go. While an undergraduate he successfully applied for research grants from the Palaeobiological Institute and the CSIRO Science Endowment Fund, easing the costs to himself. He was also supported by the Australian Skeptics Association over the three years of his science degree.

Hocknull combined his curatorial duties with a PhD at the University of New South Wales.

He has participated in the unearthing of some of Australia's most famous fossils. These include Australia's largest dinosaurs, nicknamed Cooper, George and Elliot. More recently Hocknull discovered and named three new species of dinosaur discovered near Winton, including Australia's most complete carnivorous dinosaur and two enormous plant-eating titanosaurs, one of which the carnivore may have been preying upon when it died.

Hocknull has also had the, possibly unique, honour of supervising several PhD students before he had formally finished his own doctorate, reflecting the confidence his museum and universities had in his abilities.

As Young Australian of the Year 'I did three to four presentations a week, and some of those involved three to four talks in a day … everything from primary schools to Prime Minister and Cabinet'. While the short time available in speaking to the Cabinet was not enough to deliver his message about the importance of basic science, Hocknull believes the contacts he made have given him access to the corridors of power.

'It hasn't been earthshaking, but I was a delegate to the 2020 summit, with 1000 of Australia's leading thinkers, and I've been allowed to speak on behalf of Australian science in a way that has motivated hundreds of scientists,' Hocknull says. 'It's also benefited Australian palaeontology, putting it on the radar for federal and state government departments. I hope I had some influence with the government to help fund the development of Australian natural history, like the Australian Age of Dinosaurs Museum.'

The Australian Age of Dinosaurs Museum represents the type of projects Hocknull is keen to promote. He describes the not-for-profit organisation dedicated to the preservation of Australia's fossils as 'The only place in Australia where you can come off the street and be taught to be a palaeontologist and find, excavate and prepare your own part of Australian natural history.'[1]

He gets fan mail from some of the school students he speaks to, supporting his belief that 'scientists themselves have to become the communicators' if they are going to inspire the next generation of scientists.

Hocknull is pleased that many of the groups he presents to show more interest in the palaeontology than his status as Young Australian of the Year (YAY). 'If they'd known I was out there they might have wanted me without the accolade.' The calls haven't diminished since his term finished, in part, he

1 This museum, established at Winton where many of Australia's dinosaurs are from, is still under construction.

thinks, because his replacement was tennis player Lleyton Hewitt and many people are tired of sportspeople being the spokespeople for the country. They were pleased to hear from a scientist.

Hocknull's current research focuses on recently extinct ecosystems, many traces of which are preserved in caves. After early battles between mining companies and conservationists over limestone quarrying in Queensland's caves, Hocknull is pleased to be involved in research partly funded by some of the same mining companies. This provides access and expertise in digging out any promising rocks the miners find. 'The key discoveries lead to an understanding of the processes of climate change, environmental change and faunal change[2] over the last 100 million years. It's a dream come true for someone interested in climate change and the environment,' Hocknull says.

One site offers 'The only evidence of an Australian Quaternary[3] rainforest site, covering the last 500 000 years. It's the best evidence we have of what rainforests were like before humans started mucking up the place,' Hocknull says. 'Unfortunately the results don't look so good for rainforests – they appear to have been wiped out in particular areas never to return. Around Rockhampton we see that once the ecosystems went through a phase shift they very rarely come back.'

Disturbing as his findings have been, Hocknull is excited by the significance of his subject matter. 'It's expanded our view of rainforests as connected to New Guinea. We've shared rainforest fauna for a very long period of time, only recently becoming geographically and ecologically isolated. We need to think about them together; New Guinea is part of our heritage.'

Hocknull spends more than a month per year in the field, but hopes to increase this next year. 'While I'm young I need to be out there making more discoveries for future generations.'

His work charts Australia's palaeoclimate,[4] and carries much valuable data for those wishing to discover how current ecosystems will adapt to climate change. Hocknull is very aware of the environmental significance of his work and is passionate about the topic.

'While I was YAY I had to speak at Harmony Day and other events on the topic,' he recalled. 'It was just after *Tampa*[5] and the media were keen to get me

---

2 Change in the distribution of animal species.

3 The geological era from 2.5 million ago to the present. It includes the Pleistocene and Holocene epochs.

4 Palaeoclimates are the climate conditions existing at some ancient time. Palaeoclimatology, the study of these climates, is becoming an increasingly important field of science, as knowing what past climates were like, and how animals and plants responded, gives us the best idea of what to expect in the face of rapid warming to come.

5 In 2001 the ship the MV *Tampa* collected a boatload of refugees in danger of drowning between Australia and Indonesia, setting off one of the most heated controversies in Australian history, including a huge debate about immigration and refugee policy.

to say something controversial about immigration, but I learned to be a politician and turn their questions around.'

Hocknull's preferred line was: 'Before we can live in harmony with each other we have to live in harmony with the environment because that is what is going to support us.'

Forensic archaeology: 'you have to be a bit weird'

Forensic archaeology wasn't the safest career choice for Dr Estelle Lazer, but it has certainly provided her with plenty of great stories. Lazer originally studied as an archaeologist at the University of Sydney. After completing an apprenticeship at a morgue and an anatomy course at her old University, she specialised in the little-known field of forensic archaeology.

Her first job involved emergency archaeology (studies where human or natural events may destroy the site in the near future) of burial mounds in Bahrain threatened by a causeway. The mounds were up to 5000 years old, but their importance was lost on the local goats. 'I think they must have been tubercular because they had these dreadful hacking coughs,' Lazer said. Besides spitting at her, the goats constantly tried to eat her notes, which 'had to be reconstructed around the teeth marks'.

Despite this inauspicious start, Lazer has lived two childhood dreams by working in Pompeii and Antarctica. On route home from Antarctica after a season studying the explorer Mawson's huts, she was woken for a turn as iceberg spotter. Three icebergs were visible on the radar, but their exact location was unknown. 'We were in the middle of a force 10 gale, so I had to be strapped to the mast to prevent me being washed overboard. Huge waves kept washing over the deck. It was like being in a B movie.'

There were rewards on the journey as well. 'You could see the auroras, and phosphorescent sea creatures lit up like neon signs [with] flying fish landing on the deck.'

On a trip to Heard Island she had to catalogue the sealers' huts. 'In a sort of revenge the elephant seals have taken to moulting against the sides of the huts. With a 3–5 tonne elephant seal, well you just can't budge them. Some days we couldn't get into the huts at all.' The work turned into 'rescue archaeology' as it became clear that the huts would not stand too much more punishment. Fur seals on the beach were even more dangerous, sometimes charging the team. 'We kept score some days: seals one, archaeologists zero.'

Lazer describes her work as 'a sophisticated type of voyeurism. Human bones are plastic and they are remodelled in the healing process after injuries and by activities throughout the person's life. A huge amount of information is stored in our bones.' Her work was recognised when she was named the 'Unsung Hero of Australian Science' in 2001 by the Australian Science Communicators.

Lazer found herself on the other end of human voyeurism when she worked in Pompeii, a town superbly preserved when it was buried by the 79 AD explosion of Mt Vesuvius. Most scientific attention has focused on the murals and architecture of the city, and the only previous attempts to study the bones had been scientifically dubious. Lazer established the age, height, sex and pathologies that presented on the bones of the population trapped by falling ash.

The bones were stored in ancient bathhouses with few keys. As an unknown outsider, Lazer was not popular with the local archaeologists and was locked up 'like some caged animal'. She had to bang on the doors to be let out, and didn't always get a swift response. Some tourists even threw grapes at her. However, her work has redefined ideas about Pompeii, earned her a PhD from the University of Sydney's Department of Anatomy, and led to the book *Resurrecting Pompeii*, released in 2009.

Nor has everything been easy at home. Called upon to excavate and analyse a relocated NSW cemetery, Lazer had to match the skeletons to stones before re-interring the remains. 'I spent my days lying facedown in a grave, carefully retrieving old skeletons. It was about then that I realised you have to be a bit weird to work in this field.'

As a freelance archaeologist, however, Lazer's work seldom gets boring. Besides teaching at the universities of Sydney and New South Wales, she is involved in a dig in Broadway Street in central Sydney, and spends time doing research at the morgue with the forensic dental unit.

Lazer admits there are only 'a few' forensic archaeologists in Australia, and says it is 'not a career choice if you want to end up owning your own home. It's fun, but not financially rewarding.' When Lazer applies for grants she is told her history is 'interesting', 'brave' and 'diverse' in tones that suggest these characteristics will not help her chances. However, she says, because Australia does not have enough bones of murder victims to keep anyone in regular work, most local forensic archaeologists have to maintain diverse skills.

Those expecting a world of *CSI* or *Bones* should think again. The fictional forensic scientists 'seem to do a lot more than is required of them,' Lazer says. 'I think there is more of a demarcation with the police.'

To any aspiring young scientists Lazer repeats the advice given to her, 'Keep your interests broad, and be open to opportunities.' But this has not given her much stability: 'I never know what I will be doing three weeks from now,' she says.

Things could be worse though. 'When I was young we were told if we wanted a secure career we should work in a bank.'

The iceman cometh

Dr Tom Loy achieved fame far beyond most scientists' dreams when he scored a mention in one of the biggest selling films of all time. *Jurassic Park* wasn't particularly scientifically accurate, even by the standards of Hollywood blockbusters. Nevertheless, Loy, then of the University of Queensland Molecular Archaeology Division, deserved his moment in the sun for his contribution to making archaeology both more exciting and more scientific.

Loy's research subjects included Ötzi, the 'iceman' found on the Austro-Italian border in 1991. Possibly the oldest mummified human ever found, Ötzi froze to death 5300 years ago, and was noticed by some tourists when a particularly dramatic wind from the Sahara dumped sand on the Alps, melting much of the ice as a consequence.

Scientists who have studied Ötzi have produced many competing theories as to what he was doing so high in the mountains – was he an outcast from his village, an explorer or a shepherd? Loy made detailed studies of his tools, including remains caught on them, and concluded, 'He was a specialised hunter, seeking ibex. Unlike all the other animals, ibex go up the mountains at night.' Ibex horns and skin would have had a high trading value at the time.

Loy based this conclusion on the arrows Ötzi carried, which were very long and light. 'My immediate impression was they were for long-distance shooting, not for a forested environment where they could be deflected off a leaf or twig.' Loy disagreed with some of his colleagues about Ötzi's bow, which he considered both finished and usable. 'Reconstructions show it was about a 60 pound [27 kg] bow, so pretty powerful.'

Examining what most archaeologists would consider a simple scraper, Loy found it to be a multifunction tool, 'The ancient equivalent of a Swiss army knife'. Besides scraping animal skins it had been used to cut grass, which Ötzi put in his shoes for insulation, and for separating cartilage from bone.

Loy's theories about Ötzi were confirmed when subsequent analysis of his stomach contents found the presence of ibex meat, bolstering Loy's international status.

Prior to his research on Ötzi, Loy had made a name for himself by recovering DNA from tools 1500 years old. At the time this was a shocking discovery, since it was widely believed the molecules would not survive so long. Today the idea seems laughable, as Loy pushed the oldest finds ever further back. Later Loy found usable samples in bodies buried in permafrost 20–60 000 years old. Other scientists following Loy's work have reached 800 000 years.

Loy first came to Australia in the 1980s to see how long DNA could survive under Australian conditions. While working at the Australian National University he was able to locate DNA traces 100 000 years old. It was for this work that Loy scored a mention in *Jurassic Park*, 'But I never saw a dollar from it' he said.

While his oldest research, on 2 million-year-old African tools, is nowhere near the dinosaur era, Loy suspected such finds may be possible. 'Every indication is that unless something radically changes in the soil chemistry [DNA] will last a very long time. It seems to depend on the initial conditions.'

'I'd love to be the first person to find DNA in a *Jurassic Park* scenario,' he added wistfully. However, he doubted that the idea of extracting dinosaur blood from mosquitoes trapped in amber, as in the film, would succeed, because the insects' digestive system would have destroyed the DNA before it was embalmed.

When interviewed, Loy was working on nanobes. These extraordinary objects were found by geologist Dr Philippa Uwins (then at the University of Queensland) in 1998 in ancient rocks off Western Australia. They show some indications of being alive, but are many times smaller than is generally considered the lower limit for lifeforms. The question of whether nanobes are actually alive has aroused great excitement.

In 1996, NASA scientists claimed tiny structures in a Martian meteorite were fossilised lifeforms, but received fierce criticism on the basis that nothing that small could be alive. Uwins' and Loy's nanobes are of similar size to the Martian structures, and if their living status is confirmed it would have huge implications for the search for life on Mars.

Loy was not able to synthesise nanobe DNA, but claimed, 'They eat polystyrene, as well as detergents used to kill other microbes.' He said when fed their favourite diet they will grow for around 21 days 'and form the most extraordinary Byzantine shapes.'

Loy grew up in Los Angeles. He did not begin his career as an archaeologist, and related 'At five I had a rock collection, and became hooked on geology.' Having acquired an undergraduate degree in geophysics from Redlands University, California, he moved to Alaska to work first as a seismologist, and then in petroleum prospecting. 'However, I became absolutely appalled at the rape and pillage associated with mineral exploration and decided I wanted to do something with people.' He found the 'observational science' in geology useful in his later work.

While in Alaska he studied Anthropology and Archaeology at Alaska Methodist University, but left without a degree although he had done enough to be accepted for a Masters equivalent in the same field at the University of

British Columbia. After a period working at the Australian National University he completed a PhD there in the Research School of Asian and Pacific Studies.

Dr Loy aimed to force archaeology 'to be scientifically defensible, rather than "Just So stories"'. Where previously archaeologists would say 'it looks like a scraper so it must be', now they can extract samples from an ancient tool to establish what it was used for.

Dr Loy was very proud of the course in forensic archaeology he established at the University of Queensland. 'I and one of my PhD students teach 300 undergraduate students a year, and I have 20 postgrads,' he noted. Research topics included the study of ancient genetic and infectious diseases, human and primate evolution, and identifying the remains of indigenous people in museums so they can be repatriated to their district of origin.

Update: Dr Tom Loy died of a hereditary condition in 2005, while working on a book about Ötzi. He was 63. Prior to his death he claimed to have proved that nanobes were living things, by the novel route of killing some of them. His death limited debate on whether he was right, and the question remains unsolved.

A Long history

Since 2006 Dr John Long has wowed the world with a series of astonishing palaeontological discoveries, but he was overturning theories well before that. He's now become the Vice-President of Research and Collections at the Los Angeles County Museum of Natural History.

Before leaving a post at the Museum of Victoria, he described his new position as putting him 'in charge of 35 million specimens and 80 staff'. By Australian standards the scale is awesome. The museum has the second-largest collection of any US museum, with a crucial role cataloguing the biodiversity and extinct fauna of western North America.

The main museum site is building three new galleries as part of a US$115 million redevelopment. One of these galleries will be an enormous L-shape formed from two huge halls housing some of the finest dinosaur fossils in the world. Long excitedly announces that this includes 'real *T. rex* skeletons, not replicas'. He's also in charge of one of the world's most remarkable museums, the La Brea tar pits, where ice-age animals were buried in asphalt in the heart of what is now Los Angeles. At walking distance from city hall, visitors can read about the extraordinary extinct wildlife of the area and watch palaeontologists dig mammoths and sabre tooth cats out of the sticky pits.

Long says that 'leaving Melbourne is hard,' but the appointment is something of a dream for a boy who started collecting fossils at the age of seven. The first fossil he collected was a trilobite,[1] which he later discovered was of a species that was not scientifically described until several years after he had collected his specimen.

At the age of 13 Long wrote two 100-page volumes called 'Fossils of Victoria', describing the extensive collection he had put together over the previous 6 years. He was awarded the top prize in the junior division of the Australian Science Talent Search and $60. 'Many of the species I found hadn't been described, so I described them,' Long says. Alas his specimens were isolated bones from the largest family of fish and this work was never published.

1 A class of invertebrates that dominated the oceans 540 million years ago, and died out around 250 million years ago.

Long was determined to become a palaeontologist, and studied science at the University of Melbourne. Two years into the degree he realised that Dr Jim Warren and Professor Pat Vickers-Rich (see page 2) were establishing a vibrant palaeontology program at Monash and switched, completing undergraduate, Honours and doctorate degrees there.

'I had a topic for my PhD with all the finds Warren had made at Mt Howitt,' Long says. 'I got several papers published out of it.' After a few years of gaining experience through temporary placements on 'the post-doc circuit',[2] Long settled for 15 years at the Museum of Western Australia.

While he published many papers on other topics, Long's name was increasingly associated with Western Australia's Devonian Gogo formation, from which come some of our earliest well-preserved specimens of fish. Long has been part of naming 50 new genera and species of fossil fish, along with some reptiles and a dinosaur, and has written several popular books on palaeontology, including *Feathered Dinosaurs*, *The Rise of Fishes* and *Mountains for Madness – A Scientist's Odyssey Through Antarctica.*

After moving to the Museum of Victoria in 2004 Long really hit scientific gold, including the discovery that a 380 million-year-old placoderm fish gave birth to live young rather than laying eggs. This is the oldest evidence of live birth ever found.

The story was so big *Nature* organised a live link between the Australian Science Media Centre and London, where Long got to make his announcement to 60 British journalists, members of the Royal Family, and David Attenborough. 'It doesn't get much better than that,' Long says.

'The thing I'm proudest of is publishing five *Nature* papers in three-and-a-half years,' Long says. 'Four on Gogo fishes and one on mammals from the Nullarbor caves.'[3]

The Gogo work is so important that Long remains an honorary research associate at the Museum of Victoria after changing countries, and has had the opportunity to do his own research written into the new job's contract.

Gogo is a remarkable site for the extent to which the fossils have been preserved undamaged, but Long says some very promising locations in North America are surprisingly understudied. 'There are more people working on Devonian fossils in Australia than in the whole of North America,' he says. Naturally this is something he hopes to correct.

Generally speaking, however, American palaeontology is a lot healthier than what Long left behind. 'Australia is a big country with very few palaeontologists, so it takes a long time to study what we have.'

---

2 Temporary placements for researchers with PhDs, but not enough seniority to head a team.

3 Scientific discoveries are announced through publication in peer-reviewed journals. However, not all journals are equal. Most scientists consider *Science* and *Nature* the most prestigious journals, and being accepted by either is a major achievement.

On the other hand, the United States has a culture where dinosaur fossils are hugely valued and museums compete to obtain them. Moreover, the Obama administration's 2009 budget provided a huge boost to science, raising funding by 50% in one year.

Long says that former Prime Minister Kevin Rudd's funding for infrastructure and universities has been 'good, but it hasn't helped museums'. This matters, he says, 'because we do the non-commercial work, filling in the gaps other institutions don't do, like taxonomy and understanding our biodiversity'.

Long says his role at the Los Angeles County Museum will enable him to 'build a research framework' for a major institution in just these fields. 'We'll be putting a direction on things in terms of the big questions facing society: climate change, habitat loss, biodiversity. Learning about what we have and what we could be losing.'

Astronomers and space scientists

Space suits that fit

The Apollo astronauts' bulky, rigid space suits are so much a part of our image of space travel it's hard to imagine spacefarers wearing anything else. However, Dr James Waldie is working on two sorts of spacesuits that could greatly improve the prospects of future astronauts doing their jobs well and returning healthy.

The traditional spacesuit is worn for work outside the ship, but it is so cumbersome that astronauts can't wait to get inside and into their shirt sleeves. Waldie has one suit to replace the ones used during extra vehicular activity, and another for inside wear.

There are unfortunate consequences of long stays in zero gravity. 'On Earth, our bones are strong to support and move our body mass, but in space astronauts float around without any weight or loading,' Waldie says. 'Their bodies adapt by allowing their bones to weaken at an alarming rate – it's like an extreme version of osteoporosis.' Exercise can help offset this, but astronauts can't do enough to avoid problems on return.

To solve this Waldie designed the gravity-loading skinsuit. The suit's elastics apply forces in the vertical direction that replicate the normal loading due to bodyweight when standing on Earth. The suit tricks the bones into thinking the astronaut is still on the ground, and so they stay strong and healthy. The suit cannot impose impact loads like those created when walking and running, so astronauts will still need to run on treadmills or other exercise devices to keep their bones strong.

Waldie says the suits 'feel comfortable, as if one was wearing a wetsuit' and even have the arms bare for easy of movement. This contrasts with a Russian-designed version so uncomfortable that astronauts have been reported to cut it up or refuse to wear it.

In Waldie's plans, astronauts on the space station or long-range missions will wear suits replicating Earth's gravity. Future explorers on the Moon and Mars will have modified versions to fill the gap between the host planet's gravity and what astronauts will face back home.

Waldie and two astronauts tested the suits on the 'vomit comet', the plane used by NASA to experiment with zero gravity conditions. This plane reaches enormous heights before plunging dramatically, introducing its cargo to short bursts of weightlessness.

Interestingly, for Waldie, the plane did not live up to its name – neither he nor his co-passengers experienced space sickness. Not everyone gets nauseous when first encountering zero gravity, but for all three travellers to be unaffected is rare. 'It's possible that pressure to the soles of the feet somehow helps,' Waldie says, opening up an intriguing potential side-benefit. Nausea aside, weightlessness is one of the attractions of spaceflight. Waldie says 'the suits provide loading on the body, but still allow voyagers to float freely. They could be worn as just pyjamas, or just during exercise, or for fixed amounts during the day – we are not sure yet' what will work out the best.

Waldie describes himself as 'an engineer from year one, always interested in pulling machines apart'. As a child he created a contraption from fishing lines so he could turn off his room light without leaving his bed.

After an undergraduate double degree in Aerospace Engineering and Business Administration at RMIT, Waldie read about the effects of weightlessness on bone strength and started pondering the topic. His first move was to design a suit to replace those traditionally used for extra-vehicular activities, be it in zero gravity or on the Moon. These suits have always been rigid structures filled with pressurised gas. Waldie's version became the subject of his PhD, jointly done between RMIT and the University of California, San Diego. It has a pressurised gas helmet, but for the rest of the body pressure is applied with elastics, an idea he then extended to his internal suit.

This external suit is highly flexible. It is also light, reducing the weight that must be taken into space. A tear in a traditional suit means disaster, as the whole system decompresses, taking the oxygen supply with it. In contrast, if part of one of Waldie's suits tears in space the skin immediately beneath might become bruised, but the rest of the body would be unaffected.

Funding for the gravity-loading suit has come from the European Space Agency, but after developing the internal suits as a post doc at the Massachusetts Institute of Technology Waldie has taken time off to work on pilotless aircraft at aerospace company BAE Systems. He has now returned to Australia, but continues working for BAE Systems, taking leave when funding is available to develop the suits further.

Instead of bouncing round awkwardly like the Apollo astronauts on the lunar surface, astronauts on future missions could travel comfortably for kilometres in Waldie's external suits. One feature of his design is that the suits would go with nearly normal hiking boots. Waldie notes NASA could offset the cost of future missions by selling the rights to make what will no doubt become the most famous boots since Armstrong's 'small step' – and presumably tens of millions of replicas for sale on Earth.

Asteroid wrapped

When the Earth is threatened by an asteroid, the best defence may be an idea created by University of Queensland PhD student Mary D'Souza. It will lack the pyrotechnics of a Hollywood blockbuster, but we may be safer for it.

D'Souza won the Space Generation Advisory Council's 'Move an Asteroid 2008' contest, a competition for anyone under 33 to invent a way to save the Earth from collision with the asteroid Apophis.

Apophis has a diameter of 330 metres, much smaller than the 'dinosaur-killer' but large enough to create devastating tidal waves or a major crater. In 2029 it will pass just 25 600 km from the Earth. This approach may bump it into an orbit that will cause it to hit the Earth in 2036 or later this century.

The odds of a collision are just one in 45 000, but a potential death toll in the tens of millions is keeping the danger on people's minds. Even if Apophis misses us in 2036, eventually something of a similar size will hit unless we can take defensive action.

The movies *Deep Impact* and *Armageddon* popularised the idea of blowing up interplanetary threats with nuclear weapons, but D'Souza says 'The problem with destructive solutions is you create a lot of fragments, some of which are likely to be large enough that you have to monitor them.' Instead it is much better to alter the asteroid's orbit slightly.

Many methods have been suggested, including nearby nuclear explosions, using the gravitational tug of a satellite, and even painting the asteroid white. The last idea seems strange – colour doesn't change kinetic energy! Light, however, applies a force. Sunlight bouncing off a bright surface, rather than being absorbed, creates a slightly increased force.

Planting astronauts or even robots with paint pots on the asteroid surface poses problems, so D'Souza came up with her own variation: wrap the asteroid in 3.6 tonnes of Mylar. 'By covering around 50% of it in reflecting ribbon, the enhanced radiation pressure from the Sun could push the asteroid enough to miss the Earth easily,' D'Souza says.

Apophis is unusual for an object of this size as it rotates once every 30 hours. If one end of the mylar sheet was anchored to the surface and the other spooled from a satellite, the asteroid's rotation would see it complete the wrapping itself. D'Souza says that only four space launches would be needed to get the material, the winding equipment and the spacecraft that could carry them into orbit for assembly.

Nevertheless, challenges remain. 'We need to develop the winding technology,' D'Souza says. 'You need to not rip the mylar or get it stuck. Anchors would need to be reinforced into ends of the film to avoid it tearing.

'We would also need control systems to stabilise the satellite. Otherwise it could become like a kite,' D'Souza says. 'This needs to be developed for satellite missions in general in terms of fast rotations and precision control.'

The best time for an Apophis mission would be in the early 2020s. We will have a better idea whether it is necessary in 2013 when Apophis becomes visible again and we can calculate its orbit more exactly.

D'Souza's prize was a trip to Glasgow to present her idea to two aeronautics and space conferences. She also featured in newspapers around the world. Her win was remarkable because she only found out about the competition two weeks before the end of the three-month period in which entries were open. 'The idea came from sleep deprivation, I think,' she says. 'I was a bit delirious staying up trying to come up with an idea, but it was such an interesting challenge it was worth it.'

All this occurred as D'Souza was starting her PhD on how ablation (the removal of material from a surface) affects radiation at the surface of re-entering space vehicles. 'If we can understand the heating at the surface better we can reduce the mass of the heat shield, leaving more mass for experiments,' D'Souza says.

D'Souza is using an expansion tunnel, a device for studying the behaviour of gases under high pressures and temperatures. The expansion tunnel at the University of Queensland is the most advanced in the world, enabling more accurate simulations than NASA can achieve. She is testing a new way of simulating the ablation as spacecraft are exposed to hot gases on entering the atmosphere. 'Experiments have yielded successful results so far,' she says. 'Which is quite exciting!'

Before this, D'Souza did a Masters at the University of Pisa and the French aerospace school SUPAERO. Her thesis built on her Honours degree, also at University of Queensland, on the use of nitrous oxide as the oxidant in hybrid rockets. Nitrous oxide was used on the first privately funded manned spacecraft, *Spaceship One*, but D'Souza says it has been found to be more dangerous than previously thought, and experimentation by students has been discontinued.

D'Souza says she was 'always interested in science, but didn't know what she wanted to do'. At 13 she read a book about the near-disaster of the Apollo 13 mission (see page 31) and the engineers who were forced to solve extremely demanding problems with little time or resources to get the astronauts home safely. It inspired her on a path to a Science/Engineering degree and her current thesis.

Engineering astronomy's future

Few graduates start work on the largest project in their field, but Aaron Chippendale did. His involvement will be even greater if Australia hosts the Square Kilometre Array (SKA) radio telescope.

The SKA is a truly giant project. The radio telescope will consist of a few thousand dish-shaped antennas and approximately 10 million smaller antennas, called dipoles, that are relatives of TV antennas. The dishes will be around 15 m in diameter and will work at frequencies from 500 MHz to 10 GHz; the dipoles will be tens of centimetres in size and will cover lower frequencies from approximately 70 MHz to 800 MHz. The total collecting area will be one square kilometre, hence the name. The telescopes will be arranged in a spiral pattern, with half packed closely within a 5 km radius and the rest dotted at increasing distances up to 3000 km from the main site.

Combining the power of these antennas will create an instrument 50 times more sensitive and 10 000 times faster at surveying the sky than anything existing previously. It will map the distribution of a billion galaxies, provide a better understanding of what the universe is made of and why the universe's expansion is accelerating, and test Einstein's theory of general relativity to an unprecedented degree.

The SKA will be a huge boost for the host country. Australia's short-listing as SKA host has already attracted a number of world-leading astronomers here. 'A new generation of Australian astronomers and engineers are cutting their teeth on serious and challenging SKA projects. This valuable experience will enhance Australia's capacity in scientific and technical areas that are not limited to astronomy,' Chippendale says.

High technology components will be manufactured all over the world, but the host will also experience a boost in its capacity to deliver and operate cutting-edge scientific equipment.

Originally five nations competed to host the SKA, but first the United States, then China and Argentina either dropped out or were eliminated, leaving Australia and South Africa as the final contenders, with the decision to be made in 2011 or 2012.

The chosen site needs to be well away from radio interference but safe and accessible to infrastructure, such as the high bandwidth connections needed to carry the wealth of data the SKA will provide to the world.

Chippendale is part of a team taking radio measurements at Murchison in Western Australia in an effort to demonstrate that the site is the most 'radio-quiet' of the bidders. He also works with a team developing phased array feeds that collect 30 times more information from each dish antenna than conventional feeds. 'It's an incredible experience working with the best people on exciting projects that matter,' he says.

Chippendale's undergraduate degree was in electronic engineering at Queensland University of Technology. At the end of his third year he took part in a CSIRO vacation scholarship program at the Australia Telescope National Facility.

Chippendale stayed on the CSIRO email list and, after completing his degree, saw a job advertisement seeking an engineering graduate with an interest in astronomy. 'That was my dream job summed up in one paragraph,' Chippendale says. The contacts he had established in the previous year enabled him to show CSIRO just how keen he was, and he landed the position.

CSIRO nurtured his interest and expertise by supporting his completion of a PhD in radio astronomy at the University of Sydney, for which he built his own radio telescope, while also having him contribute to the SKA bid.

Science and engineering are in Chippendale's blood, as his grandfather was an expert on soil science and plant nutrition; and his father built radio broadcasting antennas. 'My brother and I were probably among the first Aussie kids to play with a home computer. Our Dad built it from a kit and you had to type in the code,' he says.

His father had a small refracting telescope but it was missing a mount. Chippendale decided to fix this by grabbing his father's music stand and some gaffer tape. 'I used his prized telescope and stand out in the weather – but he was very supportive.'

This enthusiasm paralleled a love of finding out how mechanical things worked. 'I used to take to my toys with a screwdriver,' Chippendale says. It seems this curiosity never left him, but he says he likes even better using the things he has built to find out how the world works. 'With science there is always the possibility of finding something totally unexpected.'

The work on the SKA is an ideal combination for Chippendale. His radio monitoring data went to a panel of respected astronomers and helped establish the superiority of Australia's radio-quiet site. This, however, is not the end of the story. 'It is such a big project that astronomers won't have the final say on where to build it,' Chippendale says. A meeting of government and funding agency representatives from 19 participating nations will make the final decision, with politics playing a role along with site quality and the technical capacity of the bidders.

Even if the South African site is chosen, Australians still expect to do a lot of design work for the telescope. To make the telescope viable, the cost of highly sensitive radio antennas has to come down. Australia is at the forefront of phased array feed designs that may allow the SKA to be built with fewer antennas, saving costs in the process.

In 2009 construction commenced on the Australian SKA Pathfinder (ASKAP) near the candidate core site SKA site in Murchison. This consists of 36 dish antennas of 12 m diameter, each equipped with a phased array feed. ASKAP will provide key results to the SKA design effort and will boast a survey speed that vastly exceeds current telescopes. Perhaps 100 engineers and scientists will be based near the central site to keep everything maintained.

'Australia's main advantage is our long history in radio astronomy. We were involved from the start and have a reputation for building excellent instruments,' Chippendale says.

The SKA is described as a $2.5 billion project, but Chippendale explains that this is for the telescope and supporting infrastructure, leaving aside the cost of operation. But the pay-off will be huge, with the ability to see clearly back to some of the earliest days of the universe and observe the formation processes of the first stars and galaxies.

A cool entry

When the Huygens probe entered the atmosphere of Saturn's moon Titan in 2005 it produced stunning scientific data, but it carried more heat shielding than it probably needed.

The reason is that we don't completely understand how, and to what extent, heat is transferred to a space probe when it enters an atmosphere as different from ours as Titan's is. Therefore extra shielding was put in place to ensure the probe survived the journey. Every gram of shielding comes at the expense of a space probe's capacity to carry instruments, so the cost was in data collected.

Thanks to Dr Bianca Capra, future craft may be lighter, cheaper, and able to carry a greater payload. Between 2002 and 2005 Capra studied the effects of radiative heat transfer to spacecraft as they enter an atmosphere at great speed. Heat is transferred between two mediums in three ways: conduction, convection and radiation.[1] Understanding the mechanism of these modes is fundamental to applications from building design to aerospace engineering. 'Convective heat transfer associated with atmospheric entry is well researched,' Capra says, 'but our understanding of the radiative heat transfer mechanism, especially in atmospheres different from our own, is much less complete.' Heat radiation is significantly more complex than convection because different substances radiate differently.

Titan is of great interest to planetary scientists because its atmosphere is similar to that of a primordial Earth. Capra's work has particular importance for future missions to Titan, the only moon in the solar system with an atmosphere. Titan's atmosphere 'is high in the CN molecule, which is a very strong radiator,' Capra says. This molecule is the basis for cyanide, and fortunately is rather rare in the Earth's atmosphere.

---

1 Conduction occurs when molecules transfer thermal energy to their neighbours; convection involves the movement of hot gases or liquids to cooler locations. Thermal radiation is the electromagnetic emissions from materials, and accounts for the heat we feel from the Sun.

Capra's work has implications beyond the specialised field of deep space exploration. Radiative heat transfer may form a smaller proportion of the heat that objects encounter when entering Earth's atmosphere, but it is still a significant component. Capra's research for her PhD focused on testing the applicability of scaling laws associated with modelling radiative transfer. Such scaling laws are important to spacecraft engineers and researchers because they help to predict the level of radiation experienced by full-scale flight vehicles through the use of smaller test models encountering the same conditions. However, scaling is not always simple. Doubling the size may not double the radiative transfer. Accurately predicting the level of heat transfer, including radiative, that an entry probe will experience helps engineers design spacecraft that optimise the ratio between heat shield and scientific instrumentation.

Capra took heat transfer measurements obtained by placing scaled test models in expansion tubes that recreated specific flight conditions and compared them with actual test data gained from NASA's 1960s FIREII vehicle. Similar to the Apollo craft, FIREII was fired out of the atmosphere and observed during re-entry in the lead-up to the Moon missions. 'As a precursor to the Apollo crew return capsules, these vehicles were heavily instrumented and took extensive measurements of heat transfer (convective and radiative), radio signals and flight path information,' Capra says.

'The conductive heat transfer data has been extensively modelled and researched both numerically and experimentally. However, the radiative component does not seem to have been so extensively studied. Perhaps this is because for terrestrial entry, the radiative component is small compared with the convective.'

To measure the radiative heat transfer in the University of Queensland's ground testing facilities, Capra developed instrumentation that 'shielded the temperature sensing elements from the air flow and thus measured only the radiation component.' By comparing data collected from the FIREII vehicle with data collected on the scale models in the testing facilities, Capra investigated whether the proposed scaling laws for radiative heat transfer are correct in predicting flight radiative heat transfer.

Capra's results on measurement and scaling of radiative heat transfer were promising, and research is continuing at the Centre for Hypersonics (faster than sound travel) at the University of Queensland. Her research showed it was possible to measure radiative heat transfer in ground testing facilities and to relate this to levels experienced by real flight vehicles.

Continued research in this field could lead to a reduction in the size of heat shields used for entry into any atmosphere, particularly those where the radiative component dominates. The work may prove very useful if future missions to Mars or the Moon once again see returning astronauts entering the Earth's atmosphere at speeds of 11 km/s, as Apollo did, rather than the Space Shuttle's more stately 8 km/s.

Capra says, 'If you need a smaller heat shield to protect the vehicle during its atmospheric entry, the vehicle's weight is reduced and therefore so is the launch cost. Alternatively, larger payloads can be achieved for the same total vehicle weight, which is very good for science.'

Capra's undergraduate degree was in Mechanical and Space Engineering at the University of Queensland. During her four years as a PhD student at the Centre for Hypersonics, she was a two-time recipient of the Amelia Earhart Fellowship. Awarded by Zonta, an international organisation dedicated to advancing the status of women, the Earhart Fellowships are awarded each year to around 30 women studying aerospace or engineering across the world. Capra used both the US$6000 scholarships to 'help support her studies including the purchasing of a computer and allowing her to attend two international conferences in Germany and the United States of America' where she presented papers on her research.

Capra was awarded her PhD in 2006. She now works as a Senior Ecologically Sustainable Development (ESD) Engineer and Building Physicist with the company Bassets. Using her skills and knowledge of heat transfer, Capra improves the sustainability of our built environment. She uses computational assessment to investigate passive (non-mechanical) ways to maintain comfortable conditions inside, and therefore reduce buildings' energy consumption. She researches and consults on ways our buildings can reduce their impact on our environment, including the total energy used and greenhouse gas emissions, water use, recycling, and use of environmentally sustainable materials. Capra admits that 'although it appears to be a big departure from my PhD studies, heat transfer plays a significant role in determining how "comfortable" our indoor environment is, be it in the office, school or at home.'

A vision to Mars

If humans ever go to Mars or another planet they will be launched from the Earth riding a burst of flame from a chemical rocket. However, most of their journey will probably be conducted using plasma propulsion.

It's quite possible that the system used will be based on the work of Dr Christine Charles of the Australian National University's Research School of Physics and Engineering, who invented the helicon double layer thruster (HDLT). 'The HDLT is a beautiful piece of physics because it is so simple,' Charles said. 'It doesn't need any moving parts.'

The HDLT uses helicon plasma technology developed by Charles's team leader, Professor Rod Boswell.

In physics, plasma is the state of matter beyond gas, where atoms have become so energetic that a proportion of their electrons have separated to move independently; atoms left behind become positively charged ions. Plasma thrusters push plasma away from a space ship. By Newton's third law this means that the spaceship receives a push in the opposite direction. The strength of this push depends on how much plasma is put out multiplied by its velocity.

The double-layered nature of the plasma is observed in auroras and the solar corona, when two oppositely charged layers exist parallel to each other creating an electromagnetic field in between. Charles says this layer structure is 'quite hard to create in the lab' without the use of electrodes. Charles's invention is the first current-free double-layer plasma system produced, something which won her the Australian Institute of Physics 2009 Women in Physics Lectureship.

Helicons are low-frequency electromagnetic waves capable of travelling through pure metals, and can be used to heat the plasma to the temperatures required.

While it has yet to be seen how well it will operate in space, the HDLT has some theoretical advantages over the ion grid thrusters[1] currently used to

1 Ion grid thrusters bombard gas with energetic electrons to produce ions, which are then forced at high speed from the spaceship using an electric field.

adjust satellite orbits. While these tend to suffer from electrode erosion, the current-free HDLT has no electrodes to wear out. Ion grid thrusters also generally create positively charged beams that interfere with communications, requiring a secondary beam of electrons to be produced. The HDLT emits electrons automatically.

The HDLT can use hydrogen as a propellant, as well as more traditional elements such as methane. This could prove important, as occupied spacecraft produce hydrogen as waste. Mars missions will probably also need to use local materials rather than carrying everything on the round trip, so the ability to use diverse propellants is valuable. Charles has published a paper in the prestigious *Physical Review Letters* demonstrating that the HDLT also works with carbon dioxide, which is the main constituent of the atmospheres of Mars and Venus. This may make refuelling in space an option.

'Rockets produce huge thrust for a short amount of time, which you need to get away from the Earth's gravity,' Charles explains. 'Plasma thrusters produce orders of magnitude[2] less thrust per second but have a much greater exhaust velocity, so they use less mass over a long journey, making them fuel-efficient.'

The energy to produce and accelerate the plasma for satellites comes from solar panels, but deep space missions require so much power that the Sun just isn't bright enough in more distant parts of the solar system. Irrespective of the thruster employed, the energy source for remoter missions will be heat from the decay of a radioactive pellet.

While the idea of going to Mars generates the excitement, the money comes from the use of thrusters to move or stabilise satellites in orbit. Work has begun towards the launch in the next three years of a satellite propelled in space by the HDLT. Since 2006 the Space Plasma, Power and Propulsion group that Charles is part of has been working in collaboration with EADS-ASTRIUM, the largest aerospace company in the world, on the development of a HDLT for space use.

Growing up in France, from a very young age Charles remembers being primarily interested in 'science and music, so I studied both'. Her interest was reinforced at around 14–16 when she got to know a woman who had been a housekeeper to the Curies.[3] 'She told me all these stories, which inspired me.'

After studying applied physics at the French National Institute for Applied Sciences, Charles enrolled in a PhD at the University of Orléans in collaboration with the Australian National University. An early helicon thruster was sent to France for Charles to study expanding plasmas, but eventually she moved to Australia. She has kept up her music studies, for six years being enrolled at the ANU School of Music while continuing with physics part-time.

---

2 An order of magnitude is 10 times, or one-tenth.

3 Marie Curie was the first person to win two Nobel Prizes, one of which she shared with her husband.

Plasma research has many more down-to-Earth applications. 'The helicon plasma accelerator is used to etch and deposit materials, for example on electronics components. We're also using it for optical waveguides using SiO_2, [and] depositing platinum aggregate on carbon electrodes for fuel cells for the hydrogen economy,' Charles says. 'We'll focus on the thruster more in the next two to three years.'

However, while Mars may be a distant pinpoint in Charles's research it certainly arouses excitement, particularly in the context of the recent battery of probes studying the Red Planet and former US President George Bush's announcement of a mission there. Charles has no opinion on how likely future US administrations will be to provide the funding to implement Bush's proposal, a crucial question since almost no money has been committed so far.

'It is realistic to go to Mars provided there is the money,' she said. 'It will take decades, and there is a question of how much risk we are willing to take, how reliable we decide the systems need to be. Lots of people are very keen. We have 10 times as many students wanting to work on this. If a journey to Mars motivates young people and gets lots of people around the world working together it's a good thing.'

Behind the Moon landings

If having a ringside seat at one of the most exciting moments in human history wasn't enough, David Cooke has also been immortalised by one of Australia's most popular films, *The Dish*. Well, sort of.

The film is not an entirely accurate account of what went on at Parkes during, and prior to, Armstrong's famous step. For one thing, Cooke says there was no crucial power failure, and Parkes never lost track of where *Apollo 11* was. He says there was also no tension 'that I noticed' between the Australians and Americans at Parkes during the period it was collecting the vital signals from space.

Science was something Cooke was always interested in, and through high school he planned to study chemistry at university. However, Cooke says in the last couple of years of high school he became interested in electricity and making electronic devices.

In the end he studied to be an electrical engineer at Adelaide University, going on to work at the Salisbury weapons laboratories for over a decade. In 1967 Cooke moved to Parkes, taking up a radio receiving job, eventually becoming Officer in Charge from 1988 to 1993. At the time the team there was small (although not as small as the film portrays), with 'a couple of engineers and a couple of technicians' along with a handful of other staff. At the time of the Moon landings this was supplemented by a team of Americans.

Cooke was the senior receiver engineer at the time of *Apollo 11*. He was assigned to work with NASA to install their receivers, enabling Parkes to collect the signal from the spaceship.

While Cooke says there were no technical or personal problems, one part of the film that was accurate was the huge dust storm that blew up as the astronauts were preparing to step onto the lunar surface. The winds were so intense that normal safety procedures would have required stowing the dish,[1] but Cooke says John Bolton, the Director at the time, had been involved in the construction of the giant telescope and was able to monitor how close to tolerance the machine really was.

1 Locking it into a safety position.

Numerous myths have circulated about the role of Australian radiotelescopes in the landings. The first images of the moonwalks came not from Parkes but from Honeysuckle Creek, a fact the team who worked there are keen to emphasise. However, once the Moon rose high enough for Parkes to gain reception it was able to provide greater clarity, so for most of the mission the images the world saw came as a result of Cooke's work.

After the excitement was over, Cooke went outside to take a photograph and snapped the trailing edge of the storm disappearing into the distance. The only other image of the region's most famous meteorological event was taken from the lunar orbiter, showing a disturbance over eastern Australia. By being on site, Cooke became one of the few people to have seen the first moonwalk in sharp, clear images. The format of the transmission from *Apollo 11* to Earth was not compatible with television stations at the time, so the images were displayed on screens at the tracking stations, with conventional TV cameras pointed at the images. Not surprisingly, the vision that went around the world was much poorer quality than what Cooke and others on site got to see.

The original tapes have been lost somewhere in NASA's vast archives, and there are fears that by the time they are unearthed they will either be too degraded to use or there will no longer be machines capable of reading them.

A moment even more dramatic than the first Moon walk came later, during the *Apollo 13* mission. Parkes had not been planned to have any role in the mission, but the service module's oxygen tanks ruptured and the astronauts had to find their way home in the lunar module. It was recognised that transmissions from the lunar module would be so weak that only the world's largest receivers would be able to pick them up. The result was a frantic scramble to convert the receivers on the dish in time to pick up the transmissions needed to communicate with the astronauts. Once again Cooke was centrally involved in getting the equipment that saved the astronauts' lives up and working.

Cooke's career has had other highlights contributing to major scientific developments, albeit ones less famous than the Apollo missions. When UK scientist Jocelyn Bell observed the regular radio emissions of the first pulsar, the finding was so unexpected that explanations ranged from equipment malfunction to alien communication (her team dubbed the object LGM-1, for Little Green Men). Cooke was part of the first team to demonstrate that what Bell had found was real, and although rapidly spinning neutron stars are less headline grabbing than aliens the original discovery was considered significant enough to win a Nobel Prize.

Botanists and agricultural scientists

Top of the tree

For a botanist, it's pretty much as high as you can go; being in charge of the world's largest collection of living plants and a seed collection that includes 10% of all the plant species on the planet. Since 2007 the director of the Royal Botanic Gardens in Kew has been Professor Stephen Hopper. It's appropriate that an Australian is the first non-British director, since it was to these gardens that Joseph Banks brought the first scientific collection of Australia's flora and fauna after accompanying Captain Cook on his first voyage. The gardens hold more than seven million preserved plant specimens, contain a botanical library of 750 000 volumes and 175 000 prints, and employ 800 staff.

With the Earth going through the first mass extinction since the end of the dinosaurs, there could hardly be a more crucial time to be in charge of such a precious collection. In an editorial in the prestigious journal *Trends in Plant Science*, Hopper and colleagues wrote, 'The sound stewardship and management of plant communities and diversity is now an international imperative,' and as he makes clear, one where botanical gardens have an essential role to play.

Hopper's most ambitious project is a push for a complete inventory of living things, 250 years after Linnaeus founded the modern categorisation system for species. 'Currently, the Earth's life inventory reflects a nearly global list for birds and mammals, but not of plant species,' Hopper wrote, promoting the idea. 'And many regions on Earth, especially in the Southern Hemisphere, remain poorly surveyed. The deep sea, soil, forest canopy, and inaccessible terrain remain the least explored habitats. Perhaps only 10% of fungi are named. Estimates for most invertebrates and microorganisms, including bacteria and archaebacteria, are even lower.'

New technologies are enabling a dramatic increase in the rate with which we record life's diversity, but the project is a race against time, as species are wiped out before we know they exist.

Before he accepted the directorship, Hopper was Foundation Professor of Plant Conservation Biology at the University of Western Australia, and before that the CEO of the Botanic Gardens and Parks Authority, which is responsible for some of Perth's major botanic gardens and parks.

Kew has always appointed distinguished scientists as director. Hopper's record is impressive, even if he prefers not to classify himself with 'eminent biologists'. He has published more than 200 refereed papers, eight books, and eight more book-length monographs.

Having co-presented an episode of the ABC's natural history series *From the Heart*, Hopper is ready for Kew's communication responsibility. Hopper says he 'didn't really like the formal part' of schooling, but loved 'the bush and the beach'. Consequently, involvement in the Duke of Edinburgh's Awards and a high school teacher who taught him to 'respect the bush' had a major impact. He 'decided to find a career with outdoor involvement'.

On entering the University of Western Australia, Hopper's ambitions were towards marine biology, having been inspired by the unlikely figure of actor Lloyd Bridges, who played a scuba diver in the TV series *Sea Hunt*. However, botany offered a more enticing Honours program as it allowed the students to choose a single research topic rather than taking several compulsory units.

Hopper discovered that no one had seriously studied kangaroo paws, the state's floral emblem, and he won approval from the innovative evolutionary biologist Associate Professor Sid James. It turned out that Hopper couldn't have chosen a better place and time to study local biology. Almost one-third of the plant species indigenous to Western Australia's south-west have been discovered or scientifically described in the past 30 years, and Hopper has been at the forefront of this by describing 300 new species, mostly eucalypts and orchids.

A rare orchid can go unnoticed, but you might think eucalypts are hard to miss. Nevertheless, Hopper has described many species that previous generations did not notice, most recently a 20 m tall forest eucalypt 'growing on the top of a granite hill'. This is, Hopper suspects, 'the last really tall species to be discovered' in the region, but he fired up his students, telling them: 'If there are things that big not described, then what else is out there?'

Partly as a result of Hopper's work, the south-west of Western Australia has been recognised as one of 25 biodiversity hotspots, or locations rich in threatened species that don't exist elsewhere in the world, and one of the few in a temperate climate. He attributes the diversity of the flora and fauna in the area to the fact that it has gone for more than 250 million years without glaciation, ocean inundation, major vulcanism, or mountain building.

As long ago as 1860, Sir Joseph Dalton Hooker, second director of Kew, noted the anomaly that Western Australia has richer flora than the east coast, despite the fact that the more diverse landscape in the east might be expected to have produced the reverse. In wrestling with this conundrum, Hopper produced a theory of 'old, climatically buffered, infertile landscapes' (OCBIL). This description covers areas that have escaped the trauma of glaciation and mountain building, in the process missing out on the soil enrichment that results.

Hopper believes that in a region where glaciers have expanded and retreated across the landscape it is an advantage for plants to be able to spread their seed far and wide so they can be the first to colonise newly available territory. On the other hand, most soil in an OCBIL is poor, and Hopper suggests 'staying close to mother' is the best strategy. This leads to many plant species with a tendency to stick close together rather than trying to conquer the world, creating enormous diversity in a small area.

Hopper's OCBIL theory helps explain why conservation techniques that have achieved success elsewhere do not seem to work well in Western Australia, the Cape region of South Africa, and Venezuela's flat-topped Tepui Mountains. Elsewhere, single large reserves seem to preserve species better than large numbers of smaller parks, but in an OCBIL it can be worth 'conserving everything, no matter how small'.

The debate is important to Hopper because, he says, 'other environmental challenges are potentially reversible. Extinction is forever.'

A burning issue

Peter Christophersen is the only scientist in this book never to start a science or engineering degree, let alone finish one. However, his work proves there is more to science than what is taught at universities.

Fire is an inevitable part of the Australian environment, one which we desperately need to learn to manage. Inquiries held after devastating bushfires in southern Australia have regretfully concluded that not enough is known about Aboriginal fire management practices to recreate them to protect lives and local biodiversity.

Peter Christophersen and Sandra McGregor are making sure the same thing does not happen in northern Australia. Christophersen is a research officer with CSIRO Tropical Ecosystems and a Bunitj clan family member from northern Kakadu, while McGregor is an Umbukarla clan member from central Kakadu. They are married with four children.

Sandra's mother is a traditional elder of an area that borders Boggy Plain, a Ramsar-listed wetland[1] on the South Alligator River. Together the couple have united Western and traditional science to restore the plain to ecological health.

Neither has formal scientific qualifications. Christophersen left school after Year 11, and McGregor after Year 10. However, as well as saving the biodiversity of their part of the Kakadu World Heritage area, their scientific findings have attracted interest from as far away as Africa.

Aboriginal fire management dramatically changed Australia's ecosystems, and the environment evolved over tens of thousands of years to rely on regular burning. However, the knowledge of burning regimes was lost from southern Australia with the arrival of Europeans.

Radio shock jocks and their newspaper equivalents often like to say that more fuel reduction burns are the solution to bushfires, but there is plenty of

1 The Ramsar Convention commits governments to protect important wetlands around the world. These wetlands host migratory bird species, so the loss of one wetland damages the ecology of another thousands of kilometres away.

evidence that burning at the wrong time or in the wrong ecosystem can make future fires worse. The challenge is to work out when and where burning will help. Knowing what the traditional owners did for thousands of years would be a huge help.

In Kakadu, traditional owners have continued to use fire when hunting, particularly to open up territory or attract game. Christophersen says that the Boggy Plain area was regularly burnt until about 1993, when one of the traditional elders died. 'People burn in areas where they hunt. When someone passes on, no one hunts there for a while. There was still fire management, but not a lot around the wetlands.'

Having worked as a mechanic, contract shooter and tour boat operator, Christophersen became an electrical contractor for Parks Australia North and then a ranger for Kakadu National Park. In 2001 another ranger approached him: 'He said hot fires were racing out of Boggy Plain and threatening a bird hide and other visitor infrastructure in the area.'

Upon investigation Christophersen says that 'the land looked pretty sick. We saw very few birds and the wetland was choked with thick native grass (*Hymenachne acutiglama*).' They decided to do some burning and set up a scientific research project that would incorporate Indigenous knowledge.

Christophersen says burning of grasslands begins after big 'knock-down' storms occur in March–April. 'We burn just after those, when there is still a lot of moisture in the ground and the plants. The fires are cool and don't spread far.' This creates a mosaic of burnt areas.

Later, as the grasses dry, fires burn hotter and would spread further if the previously burnt areas didn't stop them from turning into dangerous conflagrations. By June–July the woodlands are all burnt, and burning of the wetlands begins. 'The fires are much hotter and there are strong winds, but they can't escape into the woodlands because they have already been burnt.' Burning ceases when the rains come in November–December.

When Christophersen and McGregor began their work, Boggy Plain was 60–70% covered in dense *Hymenachne* grasses. Within a year the area covered by *Hymenachne* had been reduced to 45% and what remained was generally less dense, allowing animals to move in and feed on new vegetation.

The plain now supports a diversity of vegetation types, including open water with lilies. 'This has attracted birds back,' Christophersen says. He reels off an array of species that are now common on the plain, including magpie geese, jacanas, brolgas and jabiru.

The couple have worked closely with scientists from a number of agencies, as well as teaching about 20 local indigenous children, including their own. Christophersen admits that he originally found interacting with scientists 'daunting', but says he is now much more comfortable. 'I haven't met a single scientist who wasn't very interested in what we do.'

In April the couple presented their research in the Kruger National Park in South Africa. Christophersen doubts their work is immediately applicable there, saying: 'I didn't see a lot of wetlands, but they have some similar species of grasslands'. A Kruger ranger has made a return visit to Kakadu to exchange fire knowledge.

The couple have also established an indigenous plant nursery. They have a contract to revegetate part of the Ranger uranium mine site with 8000 plants, and hope to extend the contract through to when the mine closes and much larger restoration will be required.

Science brings bread and peace

Dr Paul Fox grew up with a fascination for the biological sciences, saying he 'had a passion for anything that crept or crawled'. These days he's happy to be able to put that interest to use in what he calls 'a really practical way' – feeding people in some of the world's conflict zones.

As a research project manager at the Australian Centre for International Agricultural Research (ACIAR), Fox is in charge of projects in India, China and Bangladesh. He also has more dramatic localities to worry about in Afghanistan, East Timor and northern Iraq.

The violence in Iraq has presented a particular problem for ACIAR, and Fox and his colleagues have not been able to visit the site. 'At first I didn't think it would be possible to do much if we didn't have researchers on the ground,' Fox says. 'But frankly, working remotely has worked out far better than I imagined.'

The primary goal in Iraq has been to introduce farmers to zero tillage practices, with better crop genetics a secondary objective. Australian farmers have been able to reduce the loss of topsoil in dry years through avoiding ploughing, a technique known as zero till. This has the simultaneous benefit of increasing the amount of carbon stored in the soil.

The idea has been taken up in many parts of the world, but Fox says there has traditionally been more resistance in the Middle East, perhaps because 'ploughing has been a way of life for thousands of years'.

In order to get local farmers on board Fox has been working with scientists from the Iraqi Ministry for Agriculture and Mosul University. This has included phone and video link-ups, bringing Iraqi scientists to Australia for study tours, and meetings in northern Syria. Fox says that ACIAR was able to get several of their colleagues to the World Conference on Conservation Agriculture in India 2009, after which representatives from both countries spent time studying the way Indian farmers are adopting zero tillage agriculture.

One challenge has been to adapt zero tillage techniques to Iraqi conditions. Zero tillage requires seeding machines that are different from traditional tractors. 'The machinery used for Australian broad-acre farming is too large for Iraq, but generally that being used in India is too small,' Fox says. 'We've been developing medium-scale technology.'

Funding has been secured for four Masters and two doctoral student scholarships in Australia for members of the Iraqi team. Fox says that this is 'restoring Iraqi scientists to their rightful place in the global agricultural science community'.

Although the results are yet to be quantified, Fox says, 'Despite all the difficulties we've still made huge inroads with the farmers and convinced them to give up their traditional tillage practices.'

The benefits are being experienced beyond Iraq. Much of the machinery being used by the farmers has been built in northern Syria, where local farmers are making use of it as well.

'We see Iraq as a catalyst for the whole Middle East,' Fox says. Although he imagines the Syrian agricultural technology industry 'doesn't register on the [national] radar,' Fox believes it 'is important for local farmers' and he's excited at the skills of the local manufacturers.

Fox studied agricultural science at Adelaide University before doing a postgraduate degree at the University of Western Australia. He went to Mexico to do what was intended to be a two-year post-doc with the International Maize and Wheat Improvement Center (CIMMYT).

After 20 years with CIMMYT Fox spent a short period in private industry before moving to Canberra to work for ACIAR. Most of his time with CIMMYT was spent in Mexico, but Fox also assisted farmers in Ecuador, Peru and Bolivia. He had collected snakes and lizards as a child, but he once found himself closer than he wanted to the local varieties in Bolivia when parting a wheat crop revealed a nest of rattlesnakes.

Agricultural science in Mexico had other dangers. Once he was stopped by police who suspected the scientists of using their pick-up truck to harvest Christmas trees illegally. Fox got off lightly. However, a colleague was stopped on the suspicion of being a drug runner, and found it hard to convince police that the crops he was promoting were legal.

Fox has also led field trials of genetically modified wheat in Western Australia and helped develop the Genealogical Management System, a database designed to manage the huge amounts of information now available to crop researchers.

Reduced violence in Iraq inspires Fox and his colleagues to hope that one day they will be able to visit the area they have worked so hard for. If so, it is likely their work will have been a contributing factor to that peace. 'Food sufficiency is a very important plank in the social fabric of a country,' Fox says. 'That has been shown since the Green Revolution.'[1]

1 The 'Green Revolution' is the term given to the introduction of high yield cereal crops, particularly in the 1960s and 1970s. It is credited with preventing predicted starvation, particularly in India.

Chemists

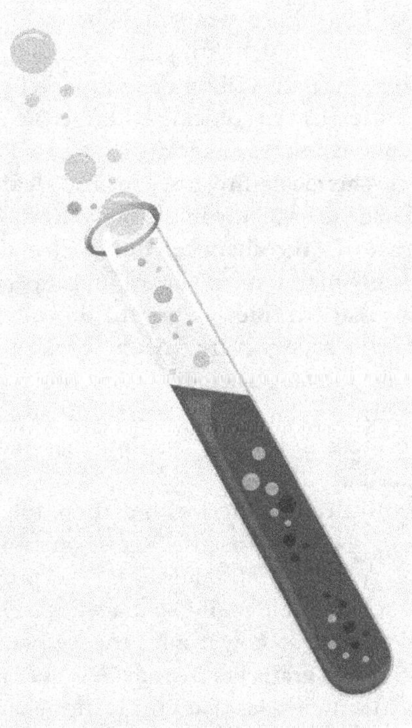

Chemistry that's better than nature

Andrea Robinson is borrowing from the extraordinary chemical diversity of the natural world to find the next generation of pain relievers, and maybe a better life for people with insulin-dependent diabetes.

Despite the pain of a cone shell sting, the complex venom produced by these invertebrates contains pain-killing chemicals. 'They use their venom to immobilise prey, and it works very quickly by targeting ion channels,[1] resulting in shutting down nerve pulses,' Associate Professor Robinson says.

Powerful as these chemicals are, they're not ideal for therapeutic use because they break down very quickly in the body. In some cases, however, we can make small changes to the chemical that will make them much more enduring, while hopefully maintaining the useful properties.

Robinson explains that enzymes within the body easily break down peptide molecules with a linear structure.[2] Many molecules resist this fate by either having ends that join together, or by having cross-chains that form cross-links. Disulfide bridges (bonds between sulfur atoms on different amino acids) are a common cross-link in nature, but these are also vulnerable to cleavage by body molecules.

'If we replace the disulfide bridges with carbon, the ring stays intact and the peptide can be more stable and useful,' Robinson says.

Other research uses similar techniques. 'Our hot project at the moment is on insulin. Insulin is not available orally, so diabetics need to inject it. We also don't know the complete story on how insulin engages with its receptor.[3] Working this out is one of the holy grails of research, because once we do that we can make mimetics [synthetic molecules that mimic the behaviour] of insulin.'

One of Robinson's contributions is to replace the disulfide bridges within the insulin molecule with more stable carbon versions. The interaction of these new insulin molecules with the body's insulin receptors offers hope of a

1 Ion channels control the voltage gradient across cell membranes, allowing cells to communicate with each other.

2 Molecules made up of amino acids (the building blocks of proteins) that are joined in a chain or line.

3 Insulin needs to lock onto a specific protein molecule, called a receptor, on a cell's surface before it can do its work of transporting glucose into the cell.

better understanding of how these receptors operate. 'Then we will see if we can stabilise insulin replacements, make them resistant to degradation and potentially deliver them orally,' Robinson says.

Robinson's PhD was on synthesising a molecule from bright orange European mushrooms, but 'My mushrooms weren't poisonous,' she says. 'I didn't go on to work in pigment research, but I've applied the skills I leant in synthesising natural molecules ever since.'

After completing her PhD, Robinson was employed by Dupont-Merck Pharmaceutical Company in Delaware, US. There she was involved in bringing pharmacological products to market. She returned to more basic research at Monash University, where she had done her Honours degree and doctorate.

In 2003 Robinson was the first recipient of the Monash Faculty of Science's 'Populate and Publish' grant, established to address the problem that scientists taking maternity leave often found their work stalled while they were away. Robinson used the money to hire a research assistant to keep her projects going while she was on leave. 'I was in daily contact with the research assistant, and she was able to run my research group, draft papers for publication and keep up with the day-to-day administration of the projects,' Robinson says. 'It meant that my research group – five PhD students, one Masters and one Honours student – were properly supervised without any drop in momentum. It also meant we were able to draft six papers for publication. I couldn't have carried anything like that workload on my own with a new baby.

'I was always a very curious child. I always wanted to know how chemicals fitted together and how chemicals interact with biology so I naturally gravitated to chemistry and pharmacology,' Robinson says. On the other hand, science was far from her only interest. She was keen on drawing and painting at school as well, and applied for fine art courses as well as science at university. 'I was always told to keep both options open. I still do art, but not in an academic context.'

Part of the reason Robinson chose a career in science was because she found it harder. 'I liked that it was challenging, liked that I learnt something new every day. But I think to do science well you have to be creative as well.'

Robinson now passes her enthusiasm for science onto another generation. She currently leads a vibrant group of 10 research students and three postdoctoral researchers. Through the School of Chemistry she also engages kindergarten students in chemistry exercises and kits kids out in lab coats and safety glasses to make slime. The students have to follow a set of instructions, and while they may not learn much about the chemical nature of their product, they love the obscene noises the slime produces.

When funding is available a more sophisticated program teaches Year 10 students about forensic science, solving fictional art forgeries, in collaboration with the National Gallery, or fictional crime scenes, in collaboration with the State Forensic Laboratory, using a collection of chemical evidence.

On the money

Almost every Australian touches David Solomon's work several times a day. You won't see his face there along with Banjo Paterson or Edith Cowan, but Professor Solomon headed the team that produced the polymers[1] used in Australia's bank notes, including the clear window designed to prevent forgeries.

Solomon's research began in the 1960s when Australian currency changed over from pounds to dollars. 'Decimal currency bank notes were state-of-the-art in resistance to forgery,' he says, 'but within a year the $10 note had been forged. It was a good forgery done with simple equipment.

'Nugget Coombs was the director of the Reserve Bank [which produces Australia's notes] at the time. He was shocked at how fast they had been forged. He was a visionary and organised a think tank.' This group of scientists spent a weekend brainstorming ways to improve note security.

'We had no suggestions at the time,' Solomon admits, but afterwards he realised that they needed to find a way to beat the photographic camera. What was required was an 'optically variable device that changed when you held the note in different light, bent it or otherwise changed the position'.

Solomon says that people were looking for evolutionary ideas, such as slightly better paper or inks. The idea of an optically variable device was revolutionary, as was his idea for a transparent section.

From there it was a small step to using film. 'The sort of film we needed was not commercially available, which was a good thing,' Solomon says, 'but we could make it by heat-laminating several films together.'

Testing their product proved difficult. Banks around the world had made such incremental changes that they had never needed to trial the durability of money.

Solomon and his colleagues ended up studying paper notes that were being withdrawn from use and creating a container that applied heat and wear to fast-track new paper notes to a similar state. They even included synthetic sweat in the mix for authenticity.

---

1 Polymers are large molecules made up of numerous repeats of identical or similar units. All plastics are polymers, as are some natural materials such as rubber and cellulose.

When put through the same test, the polymer notes passed with flying colours. Even though they are more expensive to produce than paper, Solomon says the greater durability removed the question: 'How much are you prepared to pay for security?'

'The project changed the Reserve Bank forever from an organisation that was totally dependent on overseas technology and the importation of raw materials and equipment to one that was a world leader,' Solomon says. 'That change wasn't an easy one. We were fighting to bring about a change of culture. We had to build a secret mini-production line and produce 50 million notes in order to convince the Reserve Bank that the ideas were viable.'

Polymer notes are now used in 29 countries, some of which import the plastic from Australia and do their own printing, while in other cases everything is done here.

So far the two largest potential markets, the European Union and US, have resisted making the switch. Solomon attributes this to 'inertia', since the advantages are now clear. The nonporous nature of the polymer notes even means they spread less disease.

Ironically, Solomon says he is not very interested in money, and twice left jobs to take up much lower paid, but more stimulating, positions. Solomon says he was interested in science as a child, but when he chose to do chemistry after matriculation he wasn't really sure where he was going. It was only afterwards that he grew to really love it.

At the time he was working in a paint factory and had studied the final years of school at night at Sydney Technical College. 'Chemistry made sense to me because I could relate it to the paints I was working on,' Solomon says. Solomon then went on to get a Bachelor of Science and a PhD. 'I stayed at the same place but they kept changing the name on me,' he says. Sydney Technical College is now the University of New South Wales.

Solomon's PhD was on synthetic carbonyl compounds similar to those produced by the Argentine ant. 'The ant sprays these chemicals on its enemies and we thought they might be useful as an insecticide,' Solomon says. 'They were interesting, but not the best path.'

Solomon stayed with the paint company for quite a while, only leaving when they tried to 'push me into management'. From there he went to CSIRO, where the bank note work was conducted, before moving on to the University of Melbourne. His work has brought him many of Australia's most prestigious science and engineering awards, including the 2006 Victoria Prize and the Clunies Ross Science and Technology Award in 1994. He's also a fellow of the Australian Academy of Science and the Australian Academy of Technological Sciences and Engineering, and has an honorary degree from Melbourne University, where he is still based.

While his Australian fame is mostly built on bank notes, Solomon's international reputation (which includes membership of the Royal Society) is a result of his work using free radicals to control the length of polymer chains.[2] Free radical control has allowed the tailoring of structures to the extent that single units can now be added to polymer side chains. The work has proved to have uses in everything from computer chips to bioassays testing the effect of a chemical or drug on living things.

More recently Solomon has concentrated on a project to investigate the possibility of reducing evaporation from dams by using a film one molecule thick on the top of the water. 'Evaporation is a huge problem,' he says. 'In southern Queensland they lose as much water through evaporation as they use through reticulation, so if we can make even a 10 or 15% difference it will be huge.' The idea is 60 years old but, Solomon says, 'No one has been able to get it to work.' Along with colleagues he has had success in the lab, and has just got a grant from the Victorian Government for field trials.

'I like this work because it's taking something through from idea to the market,' Solomon says. He's also working on minimising the effects of the mineral talc on the froth flotation technique the mining industry uses to concentrate the minerals they extract from ores.

Solomon likes to point out that scientists are 'learning methodologies as much as specific information', and expertise in a scientific process can enable someone to tackle problems well outside the area in which they trained. He has often been successful, he believes, because he brought a different perspective to a field in which he was unfamiliar.

2 Free radicals are chemical species that are highly reactive as a result of unpaired electrons.

Drug test leads to explosives

Not everyone is going to like Trent Pohlmann's PhD thesis. In fact, his research might make illicit drug manufacturers and aspiring terrorists pretty unhappy. Pohlmann is searching for a swift, reliable and sensitive mechanism for detecting particular chemicals. At the moment his focus is on illicit drugs, but if his technique is successful it could be used to create a test for explosives that would be suitable for screening everyone who gets on a plane or enters a secure area.

The search for such a test is widespread, but Pohlmann believes there are 'not a lot of people using our approach'. He is working on electrochemical detection, creating a selective electrode[1] suited to the particular chemical being sought.

The first target for his research is pseudoephedrine, the precursor to methamphetamine and other popular illicit amphetamines. Pohlmann is working to create a selective electrode that, when pseudoephedrine is present in solution, will cause a chemical reaction and produce an electrochemical response. He says such targeted electrodes already exist for many chemicals, but few for illegal substances.

'We need to generate something that fits like a lock and key for the chemical,' Pohlmann says. 'We also need to modify the electrode for the ideal conditions in which it operates. For example, each electrode works best at a particular pH [acidity level], which is important because sample mediums like blood and urine are quite different.'

One of the features of this technique is that the response produced is in proportion to the concentration of the drug. Some other techniques only detect a presence or absence, and can be set off by tiny levels of cross-contamination. This is just as well, since Pohlmann expects to be able to detect concentrations as low as 10^{-12} molar, or around a ten-billionth of a gram per litre, depending on the size of the molecule.

Pohlmann is studying at the CQUniversity Australia, but both Queensland University of Technology and Queensland Health Forensic and Scientific Services have assisted his work. Each new chemical to be tested will require a

1 A selective electrode transmits an electric current only in response to a particular electrochemical reaction and so can be used to detect particular chemical substances.

new sensor, but Pohlmann believes that work done on a particular drug will carry over to drugs with similar structures. When building a sensor for a chemical similar to one already developed, 'the core of the electrode should also be similar – it will probably have a similar operating range'.

Ultimately Pohlmann hopes it will be possible to build static detectors. For example, a device called an ioniser would cause molecules in the air to collect so that they could be analysed electrochemically. He believes that eventually it will be possible for people to walk through a machine like the metal detectors at airports to have their clothes checked for the presence of particular chemicals. Such a detection process would be invaluable for fighting terrorism. At the moment airports are only able to check a small sample of people boarding planes for explosive residues because the time taken for such sweeps is simply too large to allow universal testing.

Pohlmann says he was 'always interested in the technological things' as a child, and decided on a degree in science when he was at high school. During Years 11 and 12 he studied for a pilot's licence and took subjects in the Bachelor of Aviation Technology at CQUniversity through the School Links Non-Award Program. These subjects earned him a credit with the Civil Aviation Safety Authority, and he was able to count them towards his science degree when he started university full time.

Pohlmann says that TV crime shows inspired an interest in forensics, but he knew that many of his contemporaries were likewise motivated and there would not be enough jobs for all of them. For this reason he chose to study for a general science degree instead.

After an Honours degree in which he developed an electrochemical detector for copper and used it to study the highly toxic Dee River, which has been polluted by the gold and copper mine at Mount Morgan, Pohlmann found a thesis that combines his interests. Besides a clear link to forensics he hopes his work will make flying safer.

The widespread availability of drug-testing devices may raise privacy concerns for some. Pohlmann says it hadn't occurred to him that his work might be put to inappropriate uses, such as parents testing teenage children without their knowledge. 'You're the first person to raise that with me,' he said. However, on consideration he doesn't think the danger is very high. 'I don't think it will be available for individuals. I see it as more for police or Customs, although I suppose businesses could use it to test their employees.'

Computer scientists

Stone circles to computer scams and clinical notes

A program that can detect online scams from the language used won Professor Jon Patrick the 2005 Australian Computer Society Eureka Prize for Information and Communications Technology Innovation. The program is remarkable, but the paths that led Patrick to it, and his achievements on the way, are truly extraordinary.

Patrick's current work draws on varied research that has seen him acquire qualifications from five universities in as many fields. Having initially gained a degree in geodetic surveying[1] from the Royal Melbourne Institute of Technology in the 1960s, Patrick travelled to Ireland where he studied ancient stone circles, particularly exploring theories that these were used by prehistoric astronomers through alignment with the rising and setting of the Sun at particular times of the year. In the process Patrick was awarded a Masters in Science from Dublin University.

When he wanted to do a doctorate in what is known as archaeo-astronomy Patrick found that the only person willing to take him on was leading Australian computer scientist Chris Wallace. Consequently, the next step on Patrick's journey was a PhD at Monash, where, according to his CV, 'he applied the principles of Minimal Message Length encoding (MML) in a study of the geometry of prehistoric stone circles', reminiscent of Terry Pratchett's line that Stonehenge was a masterpiece of silicon chunk technology.

The work on computers, and an interest in language and coding, led to a period as an Associate Professor at Deakin University, during which time Patrick acquired a Bachelor of Science degree in Psychology from that institution. His research involved information systems for studying human behaviour, combining an understanding of natural language with computer systems to produce a powerful program for analysing sporting achievements. His relatively intuitive language programs enabled commentators to track events on

1 Geodetic surveying considers the measurement of the Earth, including its gravitational field and tidal and crustal movements.

the field so that statistics for the game could be logged and called up easily. The program was the first such system and was widely adopted by sports clubs and television stations, contributing to the recent explosion of sports statistics and making possible fantasy football leagues and their counterparts in other sports.

At this point Patrick became interested in psychology and psychoanalysis. A Diploma in Behavioural Health Psychology from La Trobe University followed, after which he worked as a psychologist running, among other things, group therapy projects for men with histories of violence.

Patrick's academic career continued to flourish and in 1998 he was appointed to the Sybase Chair of Information Systems in the Basser Department of Computer Science at the University of Sydney. Somewhere along the line he published the first substantial student grammar of Basque, a language that he says 'makes you step into a different reality'.

Patrick's Eureka Prize-winning program, Scamseek, crawls through the Internet seeking websites that are designed to hook readers into falling for phoney investments. The program uses a combination of data mining to extract patterns from data, and natural language processing (having computers interact with language written for humans) to recognise the illegal sites.

Scamseek began with an analysis of the common features of known scams, and has now developed into a 'web spider' that constantly crawls the web, recognising similar language and reporting such sites to the Australian Securities and Investment Commission (ASIC). A scan of more than 40 000 sites produced 250 that ASIC believed required action.

Patrick explains that simply leaving ASIC officers to run random searches through an ordinary search engine is not very efficient. 'If you run a Google search you even get ASIC's own site coming up.' However, Scamseek is able to detect certain features that tend to distinguish fraudulent sites from legitimate financial reporting, and passes these on for further investigation.

The websites Scamseek picks up usually involve either unregistered people trying to give financial advice, or sites that attempt to attract money to unregistered investments through means such as exaggerating the value of particular shares.

The web is global, of course, and many such attempts to fleece the unwary are based overseas. Patrick says: 'ASIC has relationships with equivalent organisations overseas. If it comes across something that is based elsewhere it will notify them.'

Patrick says that Scamseek could be expanded to detect a range of other dishonest sites. He is also interested in putting language analysis to other uses. Patrick believes language can be a powerful diagnostic tool to detect psychological ill health, linking with his work on family violence. He has published a paper analysing interviews conducted before and after therapy with men who experienced violence in their upbringing. He found that if the therapy has been successful there were changes in subliminal features of their language, such as grammatical constructions and styles of expression. For example, men who previously talked in a rule-based fashion about what people 'should and shouldn't do' subsequently shifted to more cooperative language styles.

Continuing his work with language in the health sector, since 2005 Patrick has established the Health Information Technology Research Laboratory, which has completed over 70 projects with various hospitals and health organisations in its first five years. The Laboratory has three goals for supporting doctors at the bedside:

- Use of language processing to assist with extracting highly accurate information from clinical notes
- Data and text mining to support ad hoc questions on patient records, and
- The design of clinical information systems to support language processing.

Patrick has been able to place experimental versions of these technologies in the Royal Prince Alfred Hospital, Sydney, where they are being tested for further improvement.

Computers get the joke

Dr James Curran of the University of Sydney is trying to introduce computers to humour. He is doing this by attempting to teach them natural language to the point where they can deal with the ambiguity that inspires so many of our jokes.

This won't be easy, Curran says. 'Humans and robots will be misunderstanding each other for a while yet.' As an example, Curran offers an apparently simple sentence: 'Time flies like an arrow.' We might see this as a reference to time's apparent speed, or the fact that we perceive it travelling in one direction, never to be recalled.

But if you put that sentence with an apparently similar one – 'Fruit flies like a banana' – you might suddenly imagine a species of insects called 'time flies' that eat arrows. If you're confused by this, imagine the computer's response.

The more complex a sentence gets the more interpretations are possible. One sentence Curran came across has 6×10^{23} interpretations, coincidently the number of atoms in a mole of an elemental substance (e.g. 12 grams of carbon). It's also a number so large that Curran says we couldn't store all the possible interpretations on current computers, so it was calculated using certain shortcuts. The sentence has more than 100 words, which Curran says makes it 'long, but not so long that it is exceptional in the newspapers we are using'.

Curran's work involves using statistical probability to train the computers. By finding out how words are used most commonly they can be taught which interpretations are likely. One tool is a 1.1 million word database of *Wall Street Journal* text, annotated for normal interpretation.

Another strand of Curran's research looks at analysing or parsing the grammar of a sentence so that it can be run through a search engine to answer questions. Curran says Google has already begun answering questions automatically. 'If you put into Google, "How tall is Tom Cruise?" the first link you will get back has the answer. However, if you ask "How tall is Tom Cruise in centimetres?" Google has no idea. It just treats your entry as a bag of words rather than understanding the syntax.

'Most parsers are currently painfully slow, working at a rate of one sentence per second, which is impractical for long documents. We've go the fastest parser in the world. It is about 30 times as fast as others and we soon hope to make it 100 times.'

Yet another topic Curran is trying to teach computers is lexical semantics, or recognising new words from their context. If we hear the phrase, 'The blag bit the postman,' we might suspect that a blag is a type of dog. Other phrases such as 'He walked his blag' and 'That's a big hairy blag' would confirm this. Curran says it is astonishing that we can establish a word's meaning from so few examples, since computers need to be presented with thousands of examples to see the contexts in which words are typically used.

All these lines of research have developed from Curran's PhD at the University of Edinburgh. 'I don't tend to let things go,' Curran says. Now he is supervising four doctoral students and has one who has recently submitted his thesis, a remarkable achievement for someone who finished his own thesis less than five years ago.

Curran's own undergraduate degree was a Bachelor of Science with Honours from the University of Sydney, majoring in computer science. He says he finds supervising students enormously rewarding. 'Most scientists will have far more ideas than they can execute, but when supervising you can explore far more ideas and you hear theirs as well.'

The national computer science school offers another forum for Curran to engage with students, this time talented high school computer programmers. For a week each January students are put into teams of 16 and challenged to develop a search engine from scratch, as well as build a website for a charity that is keen to reach others of similar age. In between they visit the offices of leading IT companies.

'The websites are often high quality but not so polished you would want to put them up immediately,' Curran says. However, at the end 'we give the charity a DVD and they show it to their programmers and say, "These are the kind of things 17-year-olds think 17-year-olds want." Some ideas filter through.'

A novel scientist

Dr Margaret Bearman describes her background as 'mixed', an understatement that suggests her scientific caution is winning out over artistic hyperbole.

For a scientist to have published a novel, as Bearman did with *Above the Water* (Simon and Schuster 2002), is unusual enough, but Bearman also directs short films, has a certificate in performing arts, and has had radio plays broadcast on the ABC. An unusual CV for a lecturer in medical informatics.

Bearman describes her scientific work as 'educational research, online role-playing and teaching healthcare'. She studies the way medical students use simulations to learn both 'technical content and societal context'. Bearman notes, 'Unless you take account of the context, technical considerations can fail.'

This builds on her PhD, where Bearman observed medical students learning communication skills through interaction with computer-simulated patients. While she found students could learn valuable skills in this way, there was also evidence of 'unintended side-effects'.

Bearman's parents were both scientists, and she took the maths/science stream at school. However, at 18 she 'decided my ambition was to be an actor'. She also swore never to do a PhD or become an academic like her parents. Nevertheless she enrolled in computer science at university, thinking 'this would be a high-powered form of waitressing' given acting's unreliable income. After a while, Bearman realised that she was actually interested in her science studies, and that it might be possible to sustain both sides of her life.

Since then her CV suggests Bearman has pingponged around – first completing a science degree with majors in computer science and genetics, then a Certificate of Performing Arts, then back to Honours in Computer Science, with the topic 'Using Reversive Realizability to Create a Logic Complier'. A yet-to-be-completed diploma of professional writing and editing at RMIT followed, while in 2001 she received her PhD from the Monash University Faculty of Medicine.

Bearman's employment history is similar. She has worked as a drama teacher and research assistant in community medicine at different Monash campuses, and she has lectured in medicine, while working as a freelance writer, actor and director.

The appearance of shuttling between the two fields is an illusion, Bearman stresses. In fact she has maintained both aspects all the way through. While the combination of the two is exhausting, particularly with a young child to care for as well, Bearman also finds that the two fields benefit each other. 'Writing fiction is much less all-consuming than film work. Another job keeps me stimulated,' Bearman notes.

On the other hand, her scientific work has been 'informed by my art'. She believes experience writing fiction made her PhD far more accessible and well written.

Bearman adds that in science it is important to 'Examine the impact of narrative. Story telling is a fundamental unit of the patient–physician encounter.' Her interest in the uses of narrative, stimulated by her artistic involvement, has been useful in her research.

Gone are the days when science was just a way of paying the rent. 'I would probably give up science if I could make a living from the arts, but not without regret. I see my involvement in science as ongoing.'

On the other hand the interaction of the two fields hasn't always been entirely positive. When she first enrolled in her Certificate in Performing Arts, Bearman found 'There was some friction – my commitment was questioned.' This was turned around in part by the rise of the Internet, which, Bearman found, raised the status of scientists. 'Before the Internet science was considered very daggy.'

Bearman does not believe increased respect was purely because of her computer science background. 'There was a general shift; life sciences were caught up in it as well. I'm not sure about whether the same would apply to a chemist or physicist.'

On the other hand, fellow scientists are much more supportive of her artistic endeavours. 'People in science are very interested in the arts. They hear the radio plays and come out of the woodwork. Just the other day I had a meeting with a senior academic who started off by chatting about the book.'

Many scientists confide to Bearman that they 'always wanted to write'. She encourages them, pointing to other people she knows, such as a lawyer who also has a comedy act, who manage to combine professional careers with artistic pursuits. 'They have a life as an artist, but don't have to count every penny.'

Many postgraduates, struggling to survive on scholarships and desperately seeking research grants, might find it grimly amusing to realise that, compared to those dependent on the arts, they are the ones with a safe career path and income.

Bearman admits she 'doesn't have a clear plan' for where she wants her scientific career to go. 'I'm an odd peg, not only in having the arts, but also that I'm working in medicine, but I'm not clinically trained. Consequently I have to make niches and I'm not sure where the next niche is.'

Earth scientists

An explosion of science: I blast 'em, you mine 'em

Dr Simon Michaux has the job *Mythbusters* fans want – he's spent much of his career blowing things up. He's not bad to work for, either: the electron microscope isn't working when he's interviewed for the book, so he's sent the lab technicians to the pub.

At school Michaux wanted to be a mechanical engineer, but didn't get the marks so instead he studied Applied Science at the Queensland University of Technology (QUT), majoring in physics and geology. Although this wasn't what he wanted, Michaux says it was 'exactly what I needed. I think like a scientist and not like an engineer.'

Michaux finished his undergraduate studies with a unique combination of physics, geology and chemistry. Someone must have been watching because he got a call saying: 'There's a job going that you're the only one qualified for and we really want to put in a candidate.' It wasn't a dream job to follow a driller around central Queensland and put test equipment down drill holes seeking coal deposits. 'The pay was terrible and the conditions weren't nice, but I learnt a lot,' Michaux says. Mostly he learnt to fix equipment in the field without help, and to think for himself.

'Once I did some well logging in a huge corn field. You follow the drilling platform (a large truck), which flattened all the corn, forming a path to the hole. But after 14–15 hours testing, the corn has stood up again. It was dark and I couldn't see the gate. I had to get to the next site 50 km away to be ready to start again at dawn. I thought, 'The one thing you must not do is panic,' Michaux says. 'Nothing I learnt at uni was any use in that situation.' He crashed through the corn in what he thought was the most promising direction until he hit the fence, and followed it until he found the gate. Getting lost on the job is extremely bad form, but it worked out OK in the end.

Michaux began a Masters degree in geostatistics at University of Queensland (UQ). He liked the idea of using three-dimensional mapping of drilling data to predict ore behaviour on a much finer scale, but disliked the culture. 'The field is made up of former geologists who have become mathematicians but forgotten all their geology and become tunnel-visioned. Status isn't determined by how good you are but by who can shout the loudest,' he says.

Then Michaux became a lab technician, which he considers useful now that he manages technicians. 'I know what their job is because I used to do it,' he says.

Eventually Michaux started a PhD in small-scale blasting (blowing things up). He put 50 kg lumps of rock in blast chambers to study formation of dust from the explosions. 'At the end I was all set to do a runner when this project came up,' he says. 'Most people in geology pick a field and stick to it, but I have four or five. People look at my CV and ask if I have trouble sticking to things, but this role combines all my skills.' Michaux was a strong believer in changing career path if he felt where he was working was not what he wanted.

Michaux is currently at UQ's Julius Kruttschnitt Mineral Research Centre studying the way rocks respond to crushing and grinding. Exploration programs traditionally look for the richness of a mineral ore and pay little attention to how hard it will be to get the minerals from the rocks, but this can have a serious impact on a mine's profitability.

As part of a team from three universities, Michaux is trying to change this by studying how rocks are assembled and how they will respond to various crushing techniques so that the valuable parts can be removed. 'We put rocks through their paces in destructive tests to see how much energy will be needed to liberate the valuable mineral,' Michaux says. The objective is to learn more about the rock and develop a more sophisticated model for the deposit.

The technology allows him to study much smaller samples of rock and do many more tests than was previously possible, learning not only what type of crushing will work, but why. Crushing can be done in different numbers of stages with different levels of pressure, so Michaux can help miners learn which method in the 'tool kit' best suits a particular rock. Steel crushing balls wear rapidly, so predicting the right technique saves carbon emissions during steel production – as well as money.

The research will help the industry manage financial risk and variability of production, in an environment where the deposits that remain are increasingly harder to work with much lower grades.

Michaux believes that mining as we know it is coming to an end. Ore deposits are becoming lower in grade and, with climate change and declining oil supplies making cheap energy and abundant fresh water (to wash away tailings) increasingly rare, change must come. He is keen to work in recycling, saying: 'We've put all the best metal in rubbish tips.' His love of minerals remains, slicing beautiful coasters and dice (helpfully labelled with management advice such as 'pass the buck' or 'hide under desk' instead of numbers) from the rocks he studies.

Terrorism fears have made permission to work with explosives harder to obtain, but Michaux has blown up quarries of 10–20 000 tonnes of rock, filling carefully placed blast holes with explosives. He has also tested the robustness of equipment by putting it next to explosives in the middle of large blocks of concrete.

Once, when working in the field on a large drill hole (preparation for a production blast), one of the driller's offsiders was standing too close to a blast hole when the ground collapsed under him. Michaux threw himself on the ground and grabbed his colleague. Training from school camps took over and he gripped the man's wrist rather than his hand, eventually pulling him to safety. He sees the story as evidence that the most useful parts of an education are often unexpected.

Touchdown on a cold planet

When the NASA's Mars Exploration Rovers *Spirit* and *Opportunity* touched down in January 2004, one Australian was watching more anxiously than the rest of us.

Marion Anderson was the only Australian in the team of 30 scientists who helped choose the two sites for the rovers. Prior to the rovers' arrival, just three of the 12 spacecraft to land on Mars had survived the landing and continued transmitting data for more than three days, so the choice of a safe landing site was crucial. Those choices have been spectacularly vindicated. Originally planned to operate for three months, both rovers have continued operating for more than five years. At time of writing *Opportunity* is still running, having survived numerous scares, but *Spirit* has become bogged so its solar panels cannot track the Sun and it looks likely to fail soon.

A lecturer at Monash University's School of Geosciences, Anderson participated in balancing the safety of each US$400 million rover against choosing a scientifically interesting location, although by the time she became involved the initial list of 28 sites had been substantially narrowed down.

The sites chosen were Gusev Crater and Meridiani Planum, locations now more familiar to us than some of the remoter parts of the Earth. Besides an absence of sharp rocks and cliffs which might destroy the craft on landing, the sites were chosen for relatively low levels of wind and dust, features that have helped delay the inevitable day when the rovers' solar panels become too blocked to function, leaving them to freeze during the bitter Martian nights.

Another feature the landing sites shared was evidence that they may once have been the locations for water. The Gusev site appears from space to show signs of an ancient shoreline. Meridiani, on the other hand, was known as the hematite site, as it appeared rich in this form of iron oxide. 'On Earth,' Anderson explained before the craft's arrival, 'hematite only forms in standing bodies of water, like lakes or shallow seas. So if this really is hematite it will be almost irrefutable proof Mars once had water.'

The search for water is the prime role of the two rovers. 'They're not set up to look for life,' Anderson explains. But if it can be proven that Mars once had water in liquid form on the surface of the planet, prospects for life will become far higher.

While the evidence the two plucky rovers produced has not convinced every doubter, it has certainly tilted the balance of opinion towards the belief that Mars was once wet. This has reignited interest in the planet from both the scientific community and the general public. The rovers were seldom out of the news for the first few months after their landing.

To become part of the site selection team Anderson did not need to pass any particular entry tests. 'NASA initially put out a call for all scientists to provide input' for suggested sites. Subsequently some American scientists Anderson was corresponding with invited her to the States for some field research in conditions resembling those on Mars, and suggested she take part in the site selection meeting while she was there.

This is an example of something Anderson loves about her line of research. 'The planetary science community is a great bunch of people. They are the only group of scientists I know who will treat papers from undergraduate students with the same respect as from someone who has 20 years in the field.' She attributes this to the large number of respected amateurs in astronomy. 'You've got the comet spotters who may have no formal qualifications but put in the time to do the research that no one else can do.'

One of Anderson's favourite recollections is cramming into a hotel room with many of NASA's leading scientists as they watched a transmission to discover if the *Mars Observer* had successfully gone into orbit. 'It was great to see these respected scientists jumping up and down on the beds and whooping.'

Anderson herself came to Martian research by a very roundabout route. Her previous line of work – studying the earliest evidence of life on Earth – has an obvious link to the quest for extraterrestrial life. Before that she spent 10 years as an engineer, maintaining heavy mining machinery.

In both environments being a woman made Anderson unusual. At the time she was often the only woman at geology conferences, and she would turn up to male-dominated mining camps where she found it hard to be taken seriously. At least, that was, 'until I had crawled all over the machines and got myself covered in grease.' After this the on-site engineers were willing to admit she might know something useful.

Mining machinery was never meant to be her life's work. Anderson got into the area when she found herself unemployed with a zoology degree from Monash University. 'I realised the one really saleable thing I had was a huge amount of experience with every different form of microscope. I joined a team as lab assistant and microscope operator, and when the person who headed the lab left I was the only person who really knew what was going on.'

These days Anderson mixes her work on Mars with studies of the early development of life and the geochemistry of billabongs in the Yarra Valley.

'The thing that such a diverse career has taught me is that the great skill you learn as a scientist is how to learn,' Anderson claims. 'If you realise that as a scientist you are one of the best people in the world to make yourself an expert at something, then you can do whatever you want.'

Smoother sailing

When in 2003 landlocked Switzerland became only the fourth nation in history to win the America's Cup, the world's most prestigious and expensive ocean racing event, it was a shock to many. However, it turns out the Swiss had a secret weapon – assistance from CSIRO Atmospheric Research in the form of Dr Jack Katzfey and his team.

'Challengers for the America's Cup are always looking for new technology that can help them perform better by getting the most out of the wind,' says Katzfey, the CSIRO Mesoscale Modelling Applications Team Leader. 'In our contract with the Alinghi syndicate we used CSIRO Marine and Atmospheric Research's new global weather model to forecast wind variations within the 5 km race area. We also worked with the team to develop new ways to display weather observations so they could see what the wind was doing in the first crucial minutes of a race.'

The result was a triumph for the Swiss, beating the Americans 5–1 in the challengers' round and the host New Zealanders 5–0 in the final. The outcome was also a vindication of the advice provided by Katzfey and other CSIRO scientists. In the third race of the final against New Zealand, the Alinghi team's victory margin was almost exactly the advantage they gained after receiving prior notice of a forthcoming wind shift.

Katzfey says he first became interested in meteorology 'at about 10 or 11'. He remembers looking out the window of his house in Wisconsin and seeing a dust devil. At about the same time he studied weather maps in school and his interest was fixed. 'I started producing my own barometers. I broke some tubes in the process, to my mother's distress.' (The mercury in barometers is both toxic and very hard to clean up.)

A degree in meteorology and then a doctorate followed. While working at a government laboratory in New Jersey he was advised to apply for a position at CSIRO as a regional climate modeller. In this capacity he has helped produce one of the first regional climate change simulations over Australia, modelling the influence of rising greenhouse gases from 1961 to 2100.

In 2001 the Alinghi syndicate approached CSIRO to provide weather support for the Cup challenge. Several staff members worked on the project, but Katzfey was chosen as the direct contact. He put months of work, roughly half-time, into the lead-up, followed by full-time participation in New Zealand whenever the team raced.

Katzfey was chosen for the role because, besides his extensive experience in forecasting and modelling, he grew up sailing home-made boats on Lake Michigan, and he had raced sailboats at university.

Until the challenger round was safely won the Alinghi team kept Katzfey quiet, referring to him as their 'secret weapon'. However, at the end of January 2003 Katzfey said: 'There's only Team New Zealand left and it is too late for them to do much different.' CSIRO hoped to gain valuable publicity from their contribution, and it was decided the time was right to announce it.

Although he was officially a contractor rather than a team member, Katzfey believes the syndicate was very happy with his performance. When Alinghi won the trophy he was thrown in the water along with the crew, symbolising his acceptance by the team.

'Because of the secrecy we don't know exactly what other teams had access to,' Katzfey says, but he doubts there would be anything to match the assistance Alinghi was getting. 'Our display software is unique. Other models are less advanced, have less advantageous features. The CSIRO model is state-of-the-art and very efficient.'

Alinghi successfully defended the Cup in 2007 in Valencia, Spain. Katzfey once again worked with the team, first in helping to choose the location of the races, then working closely with the team over two years leading up to the Cup.

The fact that such a success story came out of a climate change model should be a slap in the face for those who dispute the validity of computer-based predictions of climate change. However, Katzfey is unfazed by the often aggressive criticism thrown at greenhouse researchers.

'The models we have designed are to present the science as currently known. The basic physics of the atmosphere is pretty well understood. If someone has scientific criticism we will investigate that, but if they just say 'We can't believe the model because it is not perfect,' well how do you respond to that? We try to make the models as accurate as possible with our current understanding of the atmosphere. If it is not a scientific issue, but an economic one, it's the government's responsibility to respond.'

While the Cup is the most glamorous application Katzfey has worked on, his skills have also been used to provide predictions for financial markets attempting to establish prices for crop futures, which depend on seasonal climate. He's also assisted energy producers and consumers planning for peak loads, which are greatly influenced by summer temperatures.

Geoscientist shapes the world

Few people over the centuries have had the opportunity to determine the boundaries of nations – a role traditionally restricted to a handful of conquerors and the occasional diplomat and scientist.

If you think scientists looks out of place on the list, you're probably not alone. Scientists aren't famous for determining where one nation begins and another ends, but it's what Phil Symonds does. In 2002 Symonds was elected to the Commission on the Limits of the Continental Shelf, the United Nations Law of the Sea treaty body that helps determine maritime boundaries. He was re-elected in 2007.

Under the UN Law of the Sea (UNCLOS), every nation is entitled to control over the oceans and seabed within 200 nautical miles of its shore, provided its claim does not overlap with that of another nation. Seabed rights can be extended further, to the edge of the continental shelf, if the outer edge of a country's continental margin extends beyond 200 nautical miles from the shoreline.

This clause looks as if it was written with Australia in mind, as it allows us to have seabed rights over vast areas off western, southern and eastern Australia.

Establishing the exact boundaries of the continental shelf is not simple, as it depends on knowledge about the shape, depth and characteristics of the seafloor. 'More is known about the morphology of the surface of Mars or the "dark" side of the Moon than about Earth's own sea floors,' Symonds says. This is where the scientists come in.

The area a nation may claim is defined by a combination of the distance from the foot of the continental slope, the thickness of sediment on the ocean floor, distance from land, and depth of water. From the Australian perspective, the outer limit is mostly 60 nautical miles from the foot of the slope, but there are many exceptions.

Distance from land is a simple measurement, but defining where the foot of the slope lies requires detailed measurements of the shape and depth of the seafloor, and determining the thickness of sediment is not a trivial exercise.

The rules seem strange – why not have one simple measure? However, Symonds explains they evolved through many years of negotiation between those with claims and landlocked nations to ensure that it was 'possible to claim the natural prolongation of the country throughout its continental margin without taking the whole ocean'.

Within the 200 nautical mile area, nations have rights over both the seabed and the water column above it, including any fish. Extended areas only provide rights to the sea floor and anything below it, including minerals and species that live on the bottom, but not to species that swim freely in the water.

For many years Symonds' day job was preparing Australia's continental shelf submission beyond the 200 nautical mile mark, a mammoth task covering an area of 2.6 million km^2 (larger than Western Australia). In April 2008 this claim, the largest and most complex lodged to date, was confirmed by the Commission. The extended area is poorly surveyed, but parts may contain significant oil and gas deposits. These are important potential resources, although the depth of the ocean and distance from land would make drilling and exploitation a challenge.

Understandably, Symonds was not able to adjudicate on Australia's bid. This wasn't a problem since there are 21 Commission members and only seven consider each claim in detail. In turn, Symonds has been a Chairman, Vice-Chairman and member of subcommissions considering submissions by Brazil, Ireland, Norway, and a joint bid by France, Ireland, Spain and the UK, and is currently working on submissions by the UK and Japan.

Symonds and his team had to 'work frantically' to ensure Australia made the November 2004 target, initially set as the point at which the first ratifying (endorsing) nations had to make their submissions. An extension was later granted until 2009 to make it easier for developing nations to collect the data they needed, but Symonds said 'Australia and a number of other developed nations stuck with the original date to be fully consistent with UNCLOS'. Symonds also acted as an adviser to Pacific Island states on their own submissions.

Australia nominated Symonds for election in 2002 and 2007 by the states that are party to UNCLOS – 134 in 2002 and 152 in 2007. There were 24 candidates in 2002 and 26 in 2007 for the 21 places on the Commission, so Symonds needed to campaign to be elected, with support from the Australian mission to the United Nations in New York. 'The mission sent my CV around, but I still had to go over there. The country representatives still need to meet candidates face-to-face.' The choice is based on a combination of the applicant's scientific merit and international attitudes to the nominating nation.

Symonds studied science at school, choosing geology as a 'fill in – I decided I liked it,' he says, and chose to major in it at the University of Tasmania in the 1970s. 'The Bureau of Mineral Resources [now Geoscience Australia] gave me a cadetship to do Honours and I've been with them ever since.'

When Symonds started at the Bureau they placed him in marine geology, which he says was 'not an area I expected,' and sent him straight to sea. 'I loved it,' he recalls.

One aspect of the job Symonds is particularly excited by is that 'a scientist can be involved in something that involves law and politics. It uses my expertise in different ways.' Much of Symonds time is spent 'explaining technical matters to lawyers and having them explain the law to me'. Symonds received a Public Service Medal for his work in 2005, and was recently awarded honorary Doctor of Science degrees by the University of Sydney and the University of Tasmania.

There are drawbacks, however. On one early voyage in the Great Australian Bight, they had engine trouble in 'a sea I would I never have believed possible ... with waves towering over the bridge'. Symonds went below deck and packed his suitcase. He is not sure why he did this, as it would have proved an encumbrance when trying to reach a lifeboat hundreds of kilometres from shore, but the situation was new to him.

Still, Symonds concludes it has all been worth it to be 'helping to set Australia's boundaries forever'.

Both sides now

Few aspects of the weather are as obvious as clouds, but until recently the study of these visible collections of water droplets and ice has lagged behind research into clear air. Professor Graeme Stephens is out to change that, and the launch of a satellite he helped get off the ground in 2006 is enabling him to do it.

Stephens is now based at Colorado State University (CSU) where he is a University Distinguished Professor, but he grew up in Ballarat and both his undergraduate degree and PhD are from the University of Melbourne. He says he always wanted to study science at university.

'It was very different time then to enter science than it is today,' he says. 'It is very unfortunate that the value of science has been diminished over the years although perhaps the pendulum might be swinging back with science slowly being recognised as a valuable pursuit. It certainly is a lot of fun to continually learn new facts about the Earth around us and it is highly fulfilling to think that what we are learning might be put to use to better manage our environment.'

While his undergraduate degree was in 'hard-core physics', Stephens thought there were not a lot of jobs there at the time so he looked around for an area that was related but might prove more employable. He settled on meteorology and atmospheric science.

After a stint at CSIRO Atmospheric Research he went to Colorado in 1984 for what he thought would be a short-term placement. He has been there ever since.

Cloud research is an area where Australia has been a leader. Some of the work Stephens was part of in the early 1980s remains the basis of what is known about the influence of clouds on climate. Stephens says that Martin Platt, who he worked under at CSIRO, 'pioneered the use of lasers to study clouds', and during the 1970s Australia was leading the world in cloud physics research, buoyed in particular by interest in cloud seeding.

'Clouds grossly affect the greenhouse effect on the planet,' Stephens says. 'So they play a very profound role in weather and climate. Yet taken altogether they are one of the most poorly understood aspects of the climate change problem.' It's remarkable that decades after humans set foot on the Moon and landed spacecraft on other planets we know so little about what occurs inside the fluffy things that float past our windows, particularly given their importance to us.

To rectify this Stephens was part of a group formed in the 1990s with the idea of creating a meteorological supersatellite capable of studying a range of features of our atmosphere. He had written a textbook on remote sensing and was keen to take things further.

Unfortunately NASA decided it would not be allocating funding at sufficient levels, so the project was broken up into a number of smaller satellites that had to compete against each other and unrelated projects for funding approval, creating a long and frustrating road of applications and lobbying.

Amid heated competition for NASA funds, CloudSat, Stephens' primary interest, was selected. It provides an image of clouds that Stephens compares to a CAT scan. It finally got approval, along with some related ideas, including a satellite known as CALIPSO (Cloud–Aerosol Lidar and Infrared Pathfinder Satellite Observations). The two were eventually launched on the same rocket and orbit just 15 seconds apart, ensuring that they can observe areas of the atmosphere almost simultaneously.

The pair form part of a larger team of satellites, known as the A-train, flying in close formation at 705 km altitude to provide interlocking data. Stephens says, 'Some of the methodologies used by CALIPSO were also pioneered by CSIRO.' At one point in the 1990s Stephens attempted to make a combined CloudSat/CALIPSO project a centrepiece of the Australian space industry, but government support was not forthcoming for such an ambitious project.

Even with the support of NASA, the project hit a number of bumps in the road. The successful launch was actually the sixth attempt, with the previous five called off because of either technical problems or weather. Ironically, on one occasion clouds were to blame.

Now finally in space, CloudSat's Cloud Profiling Radar (CPR) makes it possible for scientists to observe not just the tops of clouds but their interiors as well, providing 1000 times the sensitivity of the radars used for weather predictions. CloudSat's 94 GHz radar has begun to reveal the hidden part of the water cycle. The science has blossomed since its launch, with the data now being used in 47 different countries.

The CPR itself is nothing new, being very similar to an instrument flown on a NASA aircraft for almost a decade beforehand, but its location in space provides an opportunity to study far more clouds than ever before.

'CloudSat allows us to peer inside and observe the processes that convert clouds to rain, for example, or mechanisms that produce hail,' Stephens told CSU's alumni magazine. He notes that precipitation is 'crucial to life as we know it', but our observations of how water vapour condenses ultimately to rain 'have been pretty crude – so far'. These new observations are now telling us that the development of rain in the real atmosphere is entirely different in character from the representations in weather and climate prediction models. There are real challenges ahead to understand how rainfall patterns will change, and the modelling tools we use for this purpose are seriously deficient.

Stephens was once an active painter, but 'didn't do much for 30 years'. When he came back to it recently it seemed logical to specialise in painting clouds 'since I'd been studying them for so long'. His paintings have received a warm response and now adorn the Cooperative Institute for Research in the Atmosphere, and have been acquired for private collections. The image of Stephens shows one of his works in the background. More works can be seen at cloudsat.atmos.colostate.edu/cloud_art. He also authored the article 'The useful pursuit of shadows' that tells the story of how paintings of clouds and the evolution of meteorology as a science built on each other during the nineteenth century.

The weather forecast is cool

The south-eastern suburbs of Melbourne host a unique project: a fully equipped weather station run with input from local schools that provides up-to-the-minute data to warn students nearby of threats such as skin cancer.

The station is the brainchild of Dan Levin, a local father of two with an interest in meteorology, who put his own money into giving children a hands-on chance to learn about the weather. Levin grew up in Johannesburg before undertaking a degree in physics and meteorology at York University, Toronto. 'I decided I wanted to have meteorology as a hobby rather than a profession,' Levin says, and when he migrated to Australia in 1990 it was to work as an IT manager for travel guide publishers Lonely Planet.

In 2004, once his children were old enough to be interested in the weather, Levin says, 'I thought I'd give them the opportunity to learn about these things, so I set up a professional station.'

Located at Levin's house, the station has professional-grade equipment, including computers capable of handling the data analysis required. The instrumentation is the same as the type used by the Bureau of Meteorology.

Nevertheless, the Bureau does not use it as part of its observation network. Levin says: 'The Bureau's network needs to be quality controlled. I've no doubt the quality is there, but it is very time-consuming to demonstrate this to qualify [as part of the network].'

Instead, Levin supplies the data to nearby Caulfield Junior College. Data is streamed from his station to the school's website, where a UV rating for the area appears on the front page. Each day certain children at the College are designated to check the website. If the site shows that the UV warning level is above 3 they scramble to raise a flag in the school grounds, telling the other pupils it is time to cover up. Levin says 3 was chosen as it is the 'community standard set by the Cancer Council.' He has considered having a second flag raised when the danger level becomes even higher, but is concerned that this might confuse the younger children.

More detailed information goes up on Levin's own site, www.elliotsweb-site.com/cws/eecl.php. As the URL suggests, Levin's son Elliot is computer-minded and helped set up the sensors for the system, as well as working on the website. His daughter Rachel provides the drawings that give the site a personal feel.

Since the site went online to the public in 2005, interest has risen to the point where approximately 300 000 hits are recorded on the website each month. Levin says this means 'The ISP keeps asking for more money.' Most of the hits seem to be from members of the general public who've discovered the page as a good place to find weather information that is more local than that supplied by the Bureau. Levin has even been asked to tell one local whether his daughter's wedding would be washed out that weekend.

A more practical use for the data has been to provide information on local wind speeds and direction for people thinking of installing windmills in their back yards. This was initially done through the Alternative Technology Association, but Levin now has many people contacting him directly.

Levin continues to add new features, including reports on soil moisture for local gardeners to assist them in knowing when to water while obeying current water restrictions. 'Currently I am looking into the feasibility of automatic SMSs to subscribers that will be alerted to UV levels requiring people to be aware of the dangers posed,' he says. He's also considering a helium 'Balloon Festival' at the local beach and UV beads that change colour with the intensity of ultraviolet light striking them.

'No studies have been done on how well the children are learning, but the school found teachers considered it of such importance that they wrote into the school's rights and responsibilities that it must continue,' Levin says.

Other schools have expressed interest in replicating Levin's system, but he says: 'When they discover how much it costs they say, "Maybe we'll just use your data".'

Glenhuntly Primary, 3 km east, is about to take up this option. Levin believes that for UV tracking his results 'could be used 20–30 km around, except over the Bay'.

Levin now works as an independent IT consultant, and thinks his project may eventually have a commercial side. 'One of the things I am working on is for putting things on websites linked to the weather station. If the UV levels are strong you might start advertising sunglasses on local websites, or umbrellas if it was raining. Primarily, however, he describes it as 'a labour of love' that comes from his passion for communicating science, particularly involving children.

Engineers

Seeing hope

Whether you want cheap glasses, electrical gadgets you can power yourself, better designs for kite surfing, fun experiments to make at home or a mechanism to shake up the whole structure of industrial design, you need Dr Saul Griffith. When he's not providing the engineering tools for a better world, he's making life more exciting and fun for anyone paying attention.

Glasses are expensive, and not just if you want designer frames. For many in the developing world the cost is simply unaffordable, and uncorrected poor eyesight limits the work they can do and sometimes endangers their lives. Lenses can be mass-produced cheaply, but an optician needs a huge supply of pre-ground lenses to be able to match any individual's needs.

Griffith doesn't wear glasses himself, but says, 'My family members did and I realised how difficult it must be without them'. When he discovered that a billion people in the world have correctable vision problems but lack affordable methods to fix them it started him on a quest.

It hasn't all been smooth sailing. In an ABC interview Griffith admitted, 'I failed on a huge number of prototypes.[1] And they all didn't work in new and interesting ways. I guess I have a terrible memory for failure, which is incredibly useful to you.'

Like Thomas Edison, who said he had 'discovered 1000 ways not to make a light bulb,' Griffith was undaunted, eventually coming up with a technique that uses a programmably adjustable mould to slash the cost of lens manufacturing.

Manufacturing costs are only half the problem. Training optometrists is prohibitively expensive for poorer nations, so Griffith invented a set of goggles equipped with a sensor to scan the user's eye and determine the prescription required.

Although Griffith grew up in Sydney and did his first degrees there, he completed subsequent qualifications in America. He spends most of his time

1 Prototypes are versions of the final product that test whether it will actually work. A prototype is usually smaller or less polished than what will eventually go on sale.

there trying to get his inventions mass-produced. He combined with his friend Neil Houghton to establish Low Cost Eyeglasses (now OptiOpia), a company dedicated to giving clear vision back to those who currently cannot afford it. The company is also looking at distributing reduced ranges of precut glasses where this is the most appropriate solution.

Meanwhile Griffith has received grants from the US National Institutes of Health for further research, and says, 'The principal focus of effort right now is in ensuring the quality and reproducibility of the devices for producing the highest quality lenses possible in the simplest machine.'

After a potentially world-changing achievement like that, many people would be happy to spend the rest of their time on their hobby, which in Griffith's case is kite-surfing. That's not Griffith's style, however. For one thing, he can't help turning out designs for better kites to surf with. These are then manufactured by his sister in Australia.

He's also trying to create something that 'looks like rope, feels like rope, but as well as acting like rope it measures how much load[2] is on itself and what state it is in'. He says that this might be useful when pushing a carrying line to its limit.

Then there are the entries in the International Bicycle Design competition, and his human-powered mechanism based on the Aboriginal 'bull-roarer' for recharging phone batteries and other portable devices. The idea was that if you run out of charge you swing the phone and charger round your head, rather than having to find a power point. This has now evolved to pulling a cord.

Wind-up cranks have allowed millions without access to electricity to listen to radios or even use mobile phones, but Griffith's pull-cord generator is more efficient, potentially expanding the range of products that can be powered this way. Potenco, the company founded to develop the technologies, estimates they could lead to 'A 1% drop in greenhouse gas emissions and a 1% bump in bottom-of-the-pyramid GDP,' which isn't a bad achievement for one person in his spare time. More information is available at www.potenco.com.

Griffith is also investigating more ambitious sources of clean energy, including kites that catch the wind at altitudes where it blows all day, offering the reliability lacked by ground-based turbines. His website www.wattzon.com allows you to compare your greenhouse emissions with others and learn ways to reduce them.

There are too many things that need to be created for even the busiest inventor to do them all alone, so Griffith is working on ways to encourage others to follow in his footsteps. To inspire the next generation of scientists and engineers he has helped start www.howtoons.org, a website of cartoons explaining how to put together everything from a drink-bottle rocket to an

2 Load is the force or weight a rope is carrying, an essential factor for kite-surfing or rock climbing.

ice-cream manufacturing kit. Griffith calls them 'tools of mass construction'. Books of howtoons are now available.

The first ideas came from Griffith and his friends. Now readers send in their own suggestions, as well as variations on the versions posted. 'It's really quite wonderful,' Griffith says. 'One thing that fascinates me is how effective the community is at spelling and grammar checking!'

Another way of promoting ideas is www.instructables.com, which Griffith describes as 'something similar to open source software, except with physical objects'. The products that various designers have helped put together include low-cost methods of cholera treatment and incubators for premature babies.

Griffith's PhD at Massachusetts Institute of Technology (MIT) was in making millions of small objects self-assemble into a complex object. He says he hopes 'to find a way to focus more attention' on this work. Prior to that he did a Masters in Engineering at MIT and a Masters in Science at the University of Sydney and an Honours degree in Materials Science at the University of New South Wales. He still teaches at MIT when time permits.

Science was a logical progression for someone who says that none of his Christmas presents survived the day without being taken apart to see how they worked. His work at MIT was bolstered by a Lego scholarship, appropriate given he still uses Lego in the inventing process. He's since won too many awards and honours to list here.

More on Saul can be found at www.saulgriffith.com.

Robots ride high

Dr Andrew Howard's fridge announces that he loves robots, which is just as well since he's helped create some of the most exciting automatons on Earth – or off it, for that matter.

Howard is a senior member of the technical staff at NASA' s Jet Propulsion Laboratory (JPL) and Adjunct Assistant Professor at the University of Southern California (USC).

Although he joined JPL after the current generation of Mars rovers had been constructed, Howard has contributed to rovers planned for NASA's Return to the Moon. He has also had a hand in numerous autonomous ground vehicles for Earth-based uses, and created open source software for robot designers.

Howard says it is much more difficult to design autonomous rovers for Earth than for Mars or the Moon. 'Plants and vegetation are much more difficult to cope with than terrain,' he says. 'On Mars we worry if a hill is small enough to drive over or [if] the soil is soft enough to bog the rover, or if a slope is too steep. All of these are relatively easy to assess with vision. On Earth we have the same problems, but find it hard to tell a shrub from a boulder.'

One program saw a multi-tonne reconnaissance vehicle affectionately known as Crusher stop in front of a field of daisies, which the vehicle interpreted as a field of hazardous obstacles. Earth-based rovers also have to deal with animals, nosy humans and water.

One of Howard's biggest challenges is to ensure that none of his vehicles pose a hazard to pedestrians or cars. On the other hand, 'The challenge on Mars is extreme efficiency,' he says. 'It's very expensive per kilogram landed on Mars. The hardware is also years out of date because we use radiation hardened components based on technology that has been very intensively tested.'

Moon rovers have to overcome a less obvious problem. 'It's very hard to dissipate heat without an atmosphere,' Howard says. Since computers are a major heat producer the projects he is working on must manage with very limited processing power.

'I can't remember ever not being a science and engineering geek,' Howard says. His first robot was built from Lego at the age of 11 or younger. He began programming at a similar age when he made his parents buy a computer.

His wife, Dr Deborah Dowling, recalls when they were undergraduates and she was miserable with flu. Howard offered to make her something to cheer her up and she suggested a duck, expecting a few yellow Lego bricks in a vaguely avian shape. 'He came back with something that walked and quacked and flapped its wings,' she says. 'I was very impressed.'

Despite this background Howard planned to be a theoretical physicist rather than an engineer. 'I found physics more difficult and therefore more challenging.'

He was particularly keen on high-energy physics, but physics in Australia was in a particularly depressed state when he was finishing his undergraduate degree at the University of Melbourne.

'So I returned to my first love,' Howard says. He completed a PhD in robotic vision at Melbourne, working on indoor robots and leading Melbourne's first Robocup team.[1]

This led Howard to USC, where he worked on multi-robot projects. 'In one we had to deploy dozens of robots to search for intruders in an empty building,' he says.

In 2005 Howard joined JPL, where he was part of the Caltech team for the Urban Grand Challenge to create a vehicle that can drive in a typical urban environment.

He also provided the vision for Big Dog, a four-legged robot intended for rough terrain, which achieved fame with the web release of a video of it recovering its footing after being kicked and after slipping on icy roads.

More recently, Howard designed the docking mechanism that will join the Dragon space capsule (to be launched on the Falcon 9 rocket) to the International Space Station.

Howard says that 'it was easy going from physics to vision research. Physics remains an excellent grounding in the maths and practical aspects of robotic design. I quite often see people who are interested in robotics with strong computer programming backgrounds. However, I look for people with a maths background because I can teach them to program, but I can't teach programmers maths.'

Although Howard is near the centre of some of the most advanced robotics work in the world, he says there are plenty of opportunities for robotics engineers in Australia. 'Australia is strong in robotics,' he says. 'Returning home is always an option.'

Yet a position with NASA is hard to pass up. Dowling says: 'The 8-year-old Andrew wouldn't forgive the 38-year-old Andrew' if he missed the chance to work on a Moon rover or the next generation of missions to Mars, which are expected to collect samples to be returned to Earth.

---

1 Robocup is a sort of robot soccer world cup in which teams from around the world produce robots that play soccer against each other or compete at more exotic games.

Clean water's 'no-tech' solution

Unclean water supplies kill almost two million children under five every year, probably more than AIDS or malaria. But Tony Flynn has helped invent something that could make a very big difference. A former potter, Flynn's interest in the firing process led to a PhD studying the properties of ceramics.

Efforts to solve problems such as unclean water are typically hindered by the inability of the developing world to pay for treatments such as filters. Often villages also lack the infrastructure to make the necessary technology themselves.

Flynn's answer is what he calls a 'no technology solution' using equal amounts of clay and organic material and a cow dung open fire. The clay and organic material are mixed and shaped into a pot and then 'you stick some straw in the dung, light it and let it burn'.

Firing the clay burns away the organic material in the mixture, leaving holes large enough for water to escape but small enough to trap pathogens (poisonous bacteria). Tea leaves and rice hulls make suitable organic material, but coffee grounds are best. 'Tea and coffee are drunk all over the developing world,' Flynn notes.

The filters 'are very simple to explain and demonstrate and can be made by anyone, anywhere,' Flynn claims. 'They don't require any Western technology. All you need is terracotta clay, a compliant cow and a match.' Cow manure is a good fuel because it provides a hot and long lasting flame. After a quick wash, even in dirty water, the filter can be ready to use within an hour of being made.

Eventually the filters become blocked with either pathogens or silt. 'If it's pure pathogens you can just put it in the fire for a few minutes until it glows red hot and it will be ready for use again,' Flynn says. Filters blocked with silt may be harder to recycle, but another one can always be made given the widespread availability of the materials.

The success of the technique depends to a substantial extent on the fact that exact temperatures are unnecessary. Clay is very stable between 800°C and a little over 1000°C, while dung fires don't go above 950°C without considerable effort. A coffee-based filter will remove more than 95% of the pathogens in water, while one made with rice or tea captures fewer microbes but processes water much faster than the 0.5 L/h that a typical pot made with coffee can manage.

The technique was originally developed for a village in East Timor that lacked clean water but had a history of pottery. Flynn is keen to know how operations are going there, but since returning to Australia he has been unable to make contact with the villagers.

After four years of what he calls 'total silence' where it seemed Flynn's great idea might never become widely used, the small aid organisation Abundant Water has started producing the filters in Laos. Flynn says 'It seems to be going ahead in leaps and bounds. I've been given the grand title of 'technical adviser'. His advice is that the filter production should be altered to suit the local conditions, using waste gases from brick kilns, rather than cow dung fires. 'Temperatures are higher within the kiln chambers than is required for the filters,' Flynn says. 'But the waste gases can produce lower temperature firing.'

Flynn never set out to be a scientist or engineer. His country school offered no science subjects other than biology, so he never took any. Nor, he says, did he feel any great fascination for science at the time. His main area of interest was Chinese history. After school, Flynn spent some time as an articled clerk for a solicitor but was never admitted as a lawyer. He resigned and went to work for *The Australian* newspaper.

He then took up pottery in the late 1970s. Eventually it became harder to make a living from pottery, so he took to teaching the art at TAFE and in a maximum-security prison.

From there Flynn moved on to teaching the theory of glazed clay at the Australian National University's School of Art. After enrolling in a Masters of Arts he found himself interested in the firing process and spent most of his time in the materials section of the Engineering Department. 'I then put up my hand for a PhD scholarship and surprised many people, particularly myself, by getting it,' he says.

While he's keen to promote his filters, Flynn is also interested in moving on to other things. He has some ideas for adapting the pots so they can clean water in areas such as Bangladesh, where the problem is arsenic, rather than bacteria. Millions of people are being slowly poisoned by the output of deep wells. Flynn says he 'needs a lateral thinking chemist' to help him on this and invites correspondence via ANU.

For his PhD Flynn investigated an unusual and novel ceramic made from the fly ash from coal-fired power plants that has 'a complete new set of properties'. The material has highly predictable permeability (ability to allow fluids like water to pass through it) when it is sintered (heated as a powder until the particles stick together) at 1000–1500°C, and unlike most ceramics is capable of surviving many rounds of cyclic fatigue (repeated rounds of stress) without failing.

The material's key features result from spherical or elliptical structures, known as cenospheres, with diameters 25–350 microns across. Flynn has found that by separating these into their appropriate sizes he was able to control the properties of the ceramic, making it useful for protecting the metal components of solar collectors, and creating materials that let through certain radiation wavelengths while shielding others five times more effectively than the same density of lead.

Dancing with the flow

Dr Fiona Sofra helps mining, oil, gas and water companies process their tailings (mine wastes) in order to reduce the environmental impact. It's an important job, as the world is filled with examples of the social effects when it's not done well – in Bouganville there was even a war as a result. Remarkably, science was a second career for Sofra, who left school after Year 11 to study dance at the Arts Educational College in London.

After completing an undergraduate degree and then a PhD in Chemical Engineering at the University of Melbourne, Sofra concluded that she was getting so much consulting work there was potential for a company. Her business, Rheological Consulting Services, now employs her full-time as director and principal consultant, as well as requiring two other engineers and drawing on the expertise of her two partners, both professors of the University. 'Since we started in 2001, the amount of work we have has grown remarkably. The studies we do are both bigger and more exciting as we get more involved in other aspects of the projects rather than just our niche area,' she says.

It is far more common for scientists to transfer to the humanities rather than the other way. Sofra explains that she won a scholarship to study dance and, at the time, 'fully intended to become a professional dancer'. However, a combination of injuries and the discovery that dance 'was not as fulfilling as I thought it would be' led her to change direction. After a year's travelling she came to the conclusion that engineering had 'a wider range of job prospects than [she had] previously thought'.

Returning to Australia, she enrolled in Year 12 at the age of 21 and continued on to university.

Many people find it difficult to return to science, particularly mathematics, after a substantial absence, but Sofra says she found it easier than before she had left. 'Perhaps it was because I knew how to study better, or because I could see a purpose to it. As you get older you get better at solving problems.'

Rheology, her speciality, is 'the study of flow and deformation of matter'. It covers everything from toothpaste emerging from a tube, through water in pipes, to the waste from tailings dams. Sofra's company has concentrated on the more viscous[1] end of the field because 'that is the way the mining industry

1 Viscous liquids are those that flow slowly, like honey, rather than easily, like water.

is going. They're trying to use less water, both for environmental reasons and because water is expensive.' The company has also contracted for the food industry, studying the behaviour of foods such as cheeses.

Some work is covered by confidentiality agreements, but Sofra says a highlight was being part of a process under which [aluminium (bauxite) miner] Alcoa's tailings dam disposal sites have 'gone from being far from ideal to being state-of-the-art, and they're still improving. They are both more environmentally friendly and more economically sustainable.'

Sofra has been able to pack plenty of travelling into the job. A memorable placement took her to Jamaica, where her flight plans were changed following riots in the capital, and she had to stay in a compound behind 15-foot razor wire with guard dogs and armed security. Goats grazed inside the mine's perimeter. She has worked in Chile, remote areas of Australia, South Africa, Canada, New Caledonia and many places she would never have seen otherwise. Another big advantage of Sofra's role is getting the chance to work with people from cultures very different from her own in trying to solve common technical challenges.

Sofra's achievements have been recognised with the awarding of a Victoria Fellowship for 2001. She particularly enjoys the diversity of the job, 'doing the management and marketing as well as the science – talking to clients every day, attending conferences to learn and share knowledge and leading training courses around the world. I'm certainly not just locked in a lab.'

She also feels very lucky to be able to have both an exciting career and a happy family life, including children. 'Industry has changed remarkably even just since we started the company. There is much more flexibility and acceptance that relationships form a big part of success – it's not all about time in the office!'

The dance helped, Sofra believes, since 'I knew I could be adaptable, and was aware of my creative side. There is huge scope to think outside the box in science and the more you do this the more likely you are to have success.'

Her advice to those interested in a similar career is: 'Just do it. You can always find a million reasons not to, but it is sometimes hard to find reasons to do something. Sometimes you just have to close your eyes and jump.'

Earth houses that don't shake down

It's rare for someone who dropped out half-way through a science degree – and never returned – to supervise final-year engineering students, but Peter Hickson's path between these improbable events may lead to safe, affordable housing in earthquake zones.

Hickson studied science in the mid 1970s, but left mid-way through his second year. 'I was a fool not to finish it because it was free,' Hickson says. 'But I had no desire to be there. I like having practical goals and I'm an impatient person. I couldn't see myself in a white coat.'

'I value the scientific knowledge and skills I received at school and university and have been applying it throughout my working life. I am constantly using observation and trial and error, research and testing to improve construction systems, methods and techniques.'

Instead of finishing university Hickson entered the building industry, and eventually moved to Shoalhaven on the south coast of New South Wales. 'I wanted to live sustainably so I built a mud brick barn followed by a mud brick home. I was a carpenter, and people around the area started using my skills to help build their earth houses,' Hickson says. He also started a business producing mud bricks for local 'downshifters' (people who have reduced their hours of work because they value a relaxed lifestyle over money).

Hickson completed a TAFE certificate in building in 1991, obtaining his builder's licence. He specialises and runs courses in earth building, a term he says covers about 18 types of building with earth materials.

In 2005, an organisation in the Philippines, keen to set up a mud brick construction industry, found his classes via the web and invited him to teach them how to get started.

'I told them I'd need to do a scoping study,' Hickson says. Eventually he found himself experimenting with local soil in Gingoog City, Mindanao, but the tropical environment didn't really suit the mud brick technique 'Mud brick developed in dry parts of the world. You're putting a lot of water into the mud, and then it needs to evaporate out. In an environment with tropical storms and lots of humidity that's very hard. Plus the bricks need covering and storing from storm events while drying, so it wasn't a practical solution.'

He decided to look to wetter climates, including Devon in England, where cob construction (named after an old word for lump) was very popular before industrialisation. He also looked at successful techniques used in seismically active places. A suitable technique had to suit the local clayey soils and other resources freely available locally.

'With cob you're mixing straw into mud, and you steadily mix in more and more straw until it is soaking up the water,' Hickson says. He decided that cob suited Filipino conditions, and combined it with bamboo as a structural re-enforcement to produce a demonstration building that he was confident could stand against both the tropical monsoon and the area's frequent earthquakes.

'The Americans left behind concrete hollow block buildings, and they withstand earthquakes,' Hickson says. 'But concrete is an advanced technology, and high-cost, high embodied energy material. People there are building with concrete blocks I can't even pick up because they fall to dust in my hands. I have no idea how they even put them on the wall.'

Cutting corners on expensive materials and lack of training and supervision has led to disaster throughout the developing world as seismic activity topples buildings that do not match specifications or achieve design strength. 'There is no need to cut corners with cob because it's basically free,' Hickson says.

Because he was more familiar with temperate climates, Hickson researched Appropriate Climate Responsive Designs for tropical conditions. He applied the simple rules: build in the shade of trees; orientate to breezes; minimise morning and afternoon exposure; generate or encourage breezes within the building.

'The prototype house works brilliantly comfort-wise as a naturally [air-] conditioned tropical home. This is what I attempt to achieve in all my houses. Even in Australia I don't believe in air-conditioning,' Hickson says. 'It's just a technological fix for bad design. In Asian low cost, medium cost homes it's not even an option.'

However, Hickson fears the wrong messages were learnt. 'When I was experimenting with mud brick there were lots of people watching me, and they've started making mud brick houses, but the climate is so wet they're having a lot of problems. I want to promote my bamboo cob, but I can't keep going there as a volunteer and funding research and development. I need someone to cover my costs and pay me to do training.'

On his return Hickson entered his work in a sustainable building competition and shared first prize. However, an engineer challenged his claims of earthquake resistance. Through his involvement with the Earth Building Association of Australia, Hickson knew engineers from the University of Technology, Sydney (UTS). One took him to a meeting of Engineers Without Borders, an aid group of students and graduates who use their skills to benefit the poorest parts of the world.

'Engineering students have to do a capstone project in their final year,' Hickson says. Two students took on building a half-scale version of half his cob house on the UTS shake table (a table which can move in ways that replicate major earthquakes) under Hickson's supervision. Then, before an audience, UTS structural engineer Professor Bijan Samali simulated a 7.6 Richter scale El Salvadoran 2001 earthquake on the shake table.

'I was nervous because it was a tougher test,' Hickson says. 'With the whole house there would be twice the bracing.' The house survived shaking at 125% of the original quake's intensity (equivalent to Richter 8.5) without structural damage. 'The building is extremely strong, and we don't fully understand why,' Samali concluded.

UTS is interested in further research, and Hickson has approached aid agencies trying to gain support for a pilot training/implementation project somewhere in Asia. He has been approached by an Indonesian architect experienced in rebuilding after the disasters in Aceh and Nias. Hickson believes he has a better way of achieving affordable, safe, healthy, comfortable, durable and sustainable housing by transferring appropriate technology and capacity building that maintains and improves resilience and self-reliance. However, he wishes he had completed a science degree as his work might gain more exposure if he was undertaking research as part of a PhD.

The science of swimming

In a society where athletes are prized above scientists, sports science is one way for researchers to get some of the action. Chris Glendenning is applying tools developed to model the flow of tsunamis and volcanoes to enable swimmers to go faster.

Fluid motion is enormously complex and very hard to model accurately. Most computational fluid dynamics (CFD) programs treat the fluid as a 3-D mesh, calculating what each part is doing.

CSIRO Mathematical and Information Sciences has taken a different approach, building on work in astrophysics to treat each molecule of water separately. This has resulted in a number of successes, including far more realistic animations of water flows for films. For his PhD, Glendenning applied these techniques to modelling a swimmer's motion through the water.

Using CFD to making swimmers go faster isn't new. Glendenning notes that swimsuit manufacturers are doing the same thing to find ways to make their products slip through the water better, to the point where some suits have been banned because they give their wearers too much advantage.

'Where our technique stands out from any other research group in the world is we can simulate interactions at the surface. That's been a big limitation for other research groups around the world and that's why you find that the research they do is restricted to the underwater stage of swimming,' Glendenning says. Ultimately this means 'their analysis is restricted to the ends of the pool, whereas we're going to be simulating what's going on along the full length of the pool'.

Rather than trying to compete with the swimsuit companies, Glendenning worked on improving swimmers' strokes. 'We want coaches to make informed changes,' he says.

Progress can be slow, however. Just modelling a few seconds worth of swimming can chew up weeks of computer processing time. Glendenning started in 2005 and is hoping his results will pay off at the London Olympics in 2012.

Glendenning combined motion capture of Australian Institute of Sport swimmers with body imaging and worked on telling the model how the skin is supposed to move. The CFD work was added later. However, there were almost immediate perks to the position, with Glendenning attending the Commonwealth Games swimming trials to meet some of the athletes and trainers and 'get exposure to the feel of a meet'.

Glendenning's work could be extended to study the motion of individual swimmers, even superimposing one swimmer's stroke onto another to see whether some motions are best for all, or if different styles suit different individuals.

'We also need to validate the computer model,' Glendenning says. 'Fluid motion is so complex [so] what we do is always an approximation, and we need to test it to see if it is close enough to give accurate predictions.'

The same programs have been used to predict the way tsunamis rise as they hit land, explaining why some areas are devastated and others come through relatively unscathed. Such work has obvious implications for disaster plans, but there may be even more lifesaving potential in studies of the causes and consequences of dam breaks.

Water is not the only fluid that can be studied this way. CSIRO has also analysed the way lava flows from volcanoes and the way landslides travel down hills.

Glendenning studied mechanical engineering as an undergraduate. 'I was torn between physics and engineering,' he says. 'I chose mechanical engineering at the University of Technology, Sydney (UTS) because we did such a range of things, including lots of science.'

Having done some CFD during studies, Glendenning decided this was the area he wanted to work in. After finishing his degree at UTS he spent a little over half a year as a design engineer in the power industry before seeing an advertisement seeking someone to complete a PhD on this topic.

'I did some swimming at school, but my sport of choice is actually archery,' Glendenning admits. However, he found sports science to his liking 'because it is so multidisciplinary'. The department is very focused on the behaviour of liquids, so he doesn't expect to gain any advantage at archery competitions by modelling air resistance.

Glendenning is never likely to get a fraction of the glory of the swimmers who will benefit from his work, but with the project funded equally by CSIRO, Monash University and the Australian Institute of Sport he gained a small portion of the money showered on sporting contests in areas where Australia leads the world.

Geneticists

Forensic consulting

The deluge of TV drama series in which forensic investigators almost always nail the criminal has led to a rush of students wanting to study forensic science at university. Carol Mayne has gone one step further, setting up her own consultancy and winning a business award in the process.

While DNA evidence has solved many crimes that were previously beyond police, and has cleared some wrongly convicted defendants, it has created problems of its own. Juries are sometimes baffled by evidence that seems incontrovertible. Mayne's slogan, however, is that DNA should stand for Do Not Assume. 'It's not as infallible as people think,' she says.

If a jury hears the odds of a person other than the defendant having the same DNA as a blood or semen stain are one in 43 trillion, it is easy to get dazzled into thinking the case is closed. What may not be revealed is the chance that the DNA was contaminated, mistested, or that a lab sample somehow got swapped, as occurred in one high-profile Victorian case. It is here that Mayne comes in.

Mayne's business, DNA Evidence Pty Ltd, assists lawyers in understanding the context in which DNA was collected and tested so they can present it to the jury. Most of her work is with defence lawyers, since prosecutors can usually get the equivalent advice from the lab that conducted the research in the first place.

Mayne says she helps the lawyer understand if they are dealing with 'a degraded sample or only a partial match'. Her work is covered by Legal Aid, where appropriate, so it's not just a service to the rich.

'It's not a case of me overturning the evidence and getting someone off who was otherwise going to be sentenced to life in prison,' Mayne says. However, by ensuring the defence lawyer is not baffled by DNA evidence, she prevents it getting more weight in a trial than it may deserve. On the other hand, at least one person who was planning to plead not guilty has changed their plea after Mayne explained to their lawyer how powerful the evidence was in that case.

A second string to Mayne's bow involves the creation of DNA-laced anti-counterfeiting paints and dyes to protect works of art or valuable brands. Her most famous client was the artist Pro Hart, but Mayne has also produced labels for wineries and Olympic memorabilia.

On request Mayne may use the appropriate DNA, such as Hart's or that from the vine on which the grapes were grown. She adds, 'There is always at least one DNA from a source only I know.' Her business has grown to the point where she now employs an assistant for the lab work.

Mayne is not the only scientist in Australia providing forensic consulting to defence lawyers, but she says for the others it is a sideline or something they are doing in retirement. As consulting is her prime business she says she is 'able to work in with court deadlines' in a way others may not.

It's an impressive career for anyone, but more remarkable considering Mayne left school after Year 10 to do a hairdressing apprenticeship. Things may have turned out very differently if the hairdresser had not gone bankrupt before Christmas that year. 'I was too cool to go back to school,' Mayne says.

Instead Mayne got a government job and finished her secondary education at night. Although it took her five years to do the last two years of school, once the qualification was gained she had recovered enough enthusiasm for study to leave her job and do a Bachelor of Biomedical Science at Griffith University.

From there Mayne went on to a postgraduate diploma in molecular genetics at the University of Queensland, with a thesis on the genes responsible for tenderness in beef. When Griffith offered a Masters in forensic science Mayne saw the opportunity to move into a field that had always interested her.

Mayne won the 2005 Queensland Smart Women – Smart State Award, and she says the award 'has been fantastic for my career. It didn't just raise my profile among scientists and lawyers, it also led to lots of schools contacting me to do speaking tours.'

Speaking to high school students, Mayne explains that, 'Once you discover something you are interested in, learning becomes fun.' Mayne says she enjoyed science and maths at school, but 'sport was more fun' and does not remember ever wanting to be a scientist.

Mayne is an advocate for lifelong learning and for inventing your own career path. 'When I had done forensics I didn't want to start again at the bottom so I did the market research and found there was an opportunity and I went for it.'

Swat the difference

When Dr Steve Chenoweth talks about his work its not the kind of discussion you expect from a former investment banker. 'Differences between males and females make up a substantial component of diversity in the biological world, with the sexes often differing in size, shape and colour', he says. 'The catch from a genetic standpoint is that males and females share almost all of their genes. Because of this, many genes that benefit one sex may actually be harmful to the other.'

Arguably, it's a lot more interesting than investment banking. Maybe that is why Steve Chenoweth chose to leave his financial career behind and pursue a study of the sex differences in fruit flies.

'In birds, a gene that causes brightly coloured plumage in males may have advantages in terms of attracting a mate,' Chenoweth continues, 'whereas its effect in a female could be a distinct disadvantage, making her more noticeable to predators, for example. These so-called "sexually antagonistic" genes are a real problem, and how species have come to deal with their detrimental effects while maintaining their benefits remains a mystery for modern genetics.'

The study of these variations has taken Chenoweth to the rainforests of northern Queensland, where he collects impregnated (fertilised female) *Drosophila serrata* fruit flies to breed in his lab at the University of Queensland's School of Integrative Biology.

Chenoweth says that he chose *D. serrata* over its far more heavily studied relative *D. melanogaster*[1] because 'in melanogaster the sexes have evolved so far apart that some of the traits in males don't exist at all in females. In *D. serrata* they're still there in both sexes but are unequally developed in males and females. What is more, there is a geographic variation in these characteristics, which means that natural selection may be acting on them.'

Chenoweth says what he loves about science is 'getting to ask your own questions'. It's something he didn't get to do during a two-year interlude working as an investment banker, first at a large bank and then in a small financial institution.

---

1 The common fruit fly, *Drosophila melanogaster*, is one of the most studied species in the world, the basis for a large portion of our knowledge about genetics.

While banking had plenty to offer, particularly in terms of salary, Chenoweth says he now 'wouldn't look back'. This wasn't his reaction initially, however. 'There is an insane amount of repetition in fly work,' Chenoweth says. 'You can spend 16 hours a day doing nothing but counting flies. We sometimes have a bunch of us doing this for days and I am sure it could be used as a form of torture. You just go crazy.'

Six weeks after returning to science Chenoweth was doing this sort of boring work and opening his pay slip to find it was '90% lower' than at the bank. His scientific colleagues also thought he was mad, saying: 'Nobody comes back [to science after leaving]'.

The other thing Chenoweth found difficult about the career change was the different speeds at which the two fields operate. 'The academic lifestyle is about excellence,' Chenoweth says, 'about doing the best you can. The commercial world is about timeliness. There is not much patience for sticking at something. It's very opportunistic.'

Chenoweth says he 'always expected to be a biologist', having been interested in biology all through school. At university his undergraduate biology degree was followed by Honours and a doctorate studying the genetics of estuarine fish such as barramundi, all in the School of Environmental Sciences at Griffith University. At the end of this he found himself disillusioned with a scientific career and, after checking out prestigious laboratories at Harvard and the Smithsonian, decided to make the shift.

'I had analytical and computing skills which I leveraged into a banking career,' Chenoweth says. He found himself working with a lot of other former scientists who had also found their skills in high demand. Before the career shift, however, he had done a side project on fruit flies. The colleague he was working with at the time had continued with this research and found a very interesting pattern of natural variation, so when Chenoweth decided he wanted to get back to science he decided that fruit flies would be his species.

The sexual characteristics of social insects such as bees and ants are so different from those of vertebrates that one might question how well lessons learnt in this area can be applied. However, Chenoweth says that most Drosophila share with humans a sex determination system where females have two copies of the X chromosome whereas males have one copy of each of the X and Y chromosomes. This contrasts with sex determination in many other insects and even non-mammalian vertebrates. He adds that while the specific genes may be different in humans and flies, the mechanisms by which genes affect sexual dimorphism may be similar.

When first interviewed Chenoweth was about to go on paternity leave; his first child was born the day after. However, there was no doubt where he would be going when his leave was over. 'I couldn't wish for a better job,' he says. This view was probably reinforced when he received a University of Queensland Foundation Research Excellence award.

Thoroughbred geneticist

Associate Professor Ann Trezise wasn't interested in science at school. In fact, she failed it in Year 10. Nor was she interested in horses. The idea that she might eventually head the Australian Equine Genetics Research Centre was certainly not on her mind, but a single class transformed her life.

'In the second half of final year, my biology teacher told us about Mendelian genetics,' Trezise says. Gregor Mendel discovered that certain traits in plants and animals are determined by a gene inherited from each parent, and that certain traits were recessive so they would only show up if an individual had both genes for that trait. 'I understood it instantly, it absolutely clicked,' Tresize continues. 'I knew then I wanted to learn everything I could about genes and how they worked.' It's a fascination that has never left her.

The previously bored student dropped her idea of becoming a kindergarten teacher and decided to study science at university, but 'I'd done all the wrong subjects – things like music and history.' She repeated final year, this time studying maths and sciences, and the next year was accepted into Griffith University.

Throughout her degree in biology, Trezise's enthusiasm for genetics never waned, and she completed a PhD in the field, including early work on cloning mammalian genes. Trezise then moved to Canada and, despite shuddering at the memory of three winters in Toronto, says she had a 'wonderful time' cloning the gene for the heritable disease cystic fibrosis.

From Toronto Trezise moved to Oxford, where she worked on ABC transporter genes, a superfamily that includes the genes for cystic fibrosis and multidrug resistance in tumours. However, throughout this time she struggled with back problems, to the point that her spine has been surgically fused 'from the neck all the way down'.

The cold weather made this particularly painful, so in the mid-1990s she headed home to Brisbane. As her back improved Trezise was able to take a full-time position at Griffith before shifting to the University of Queensland in 2000. She is still based in the Faculty of Biological and Chemical Sciences.

In the course of her work in molecular genetics she was asked to act as a consultant to the Australian Equine Genetics Research Centre. The primary role of the centre is to keep track of every thoroughbred racehorse listed in the

Australian Stud Book, along with a number of other horse breeds. Up to 19 000 new foals each year must be DNA-tested by the Centre to confirm their parentage and record their identification for the Australian Stud Book. When horses are imported into Australia, the Centre tests them to confirm that the correct horse has been sent. Trezise says that in 2005 'we found a mistake. The wrong horse had been sent from overseas.' She believes this was an honest error, but says the event 'demonstrates why we need this service. The buyer wouldn't have realised he had the wrong horse otherwise.'

In February 2006, when the then director of the Centre retired, Trezise was invited to become director. With a registry of DNA from more than 180 000 horses, Trezise believes the Centre can expand its role considerably.

At the moment the Centre does not provide advice on the genetic characteristics that make a good racehorse. 'We test for 12 areas of the horse genome,' she explains. 'All are repeated units that sit between the functional genes [genes whose code produces proteins]. Those studies that use markers to try to link to characteristics require thousands of marker sites to get the density to look for co-inheritance.'

However, Trezise does believe that the Centre can help the horse breeding industry in other ways, such as through the study of genetic diseases. She gives the example of overo-lethal white foal disease, an example of the Mendelian genetics that fascinated her at school. A foal with a single copy of the relevant gene will have a white patterning that many breeders consider attractive. However, two copies of the same gene ensure the foal will die young.

The release of the fully sequenced horse genome early in 2010 gives the Centre the opportunity to investigate more widely. 'In the past, everything I have done has been basic research,' Trezise says. 'This is a very different approach. I find it an absolute joy to interact with the business people who are joint proprietors of the Australian Stud Book, the Victoria Racing Club and the Australian Jockey Club, and to report to them on the applications of my work and the areas we can move into. They are very supportive.'

Marine biologists

Sometimes cold, always beautiful

Anyone who wants to do research in some of the most beautiful places on Earth can take heart from Professor Peter Ralph. Twice each year he studies the health of coral reefs around Heron Island in the Great Barrier Reef, and once a year he heads for the breathtaking snow and extraordinary skies of Antarctica.

The experience of going from –20°C in November to +30°C in January, with only a short time in Sydney to acclimatise, is something of a challenge, but Ralph considers himself 'very lucky to work in such pristine and beautiful environments'.

Ralph's subjects are the microscopic algae and phytoplankton[1] that form the base of both ecosystems. He considers it an 'honour' to study species that hold up the rest of the food chain, particularly at a time when both coral reefs and Antarctica are under great threat.

Ralph leads a team studying coral bleaching on the Great Barrier Reef. When corals become stressed they expel the algae that give them both their energy and their colour. Without the brightly coloured algae, the corals are white or 'bleached'.

Coral can recover and repatriate the algae if the stress is mild or temporary. However, incidences of bleaching sometimes lead to massive irreversible damage to coral populations. High temperatures are among the most common sources of coral stress.

In Antarctica, Ralph studies the microalgae that live on the bottom surface of the sea-ice. The microalgae are grazed upon by krill, the food source for baleen whales and often seals and penguins.

Ralph explains that the unifying theme of his studies is that 'both ecosystems are threatened by climate change in our lifetimes, and we have very little time to learn about them'.

Ralph planned to be a scientist throughout high school, being 'in love with the chemistry lab'. However, he had no particular interest in microalgae. After an undergraduate degree in Environmental Biology from the University of Technology, Sydney (where he now works), he did his PhD at the same university on photosynthesis in seagrasses.

1 Tiny plants and algae that rely on sunlight for energy.

Despite Australia being the world centre for seagrass biodiversity, Ralph says there is very little funding for research here. Students of seagrasses have to relocate or be creative and pursue research in other fields. Ralph took the latter option, using his understanding of photosynthesis to move first to studying corals reefs and then to investigating sea-ice.

The connection between global warming and coral reef damage has received fairly wide public attention, but the danger that climate change poses to polar marine ecosystems is not as well known. It would be wrong to assume ecosystems will flourish as greater warming brings temperatures around the coast of Antarctica closer to the global norm. Ralph explains that the underside of sea-ice forms a surface on which phytoplankton can live. 'Plankton in the area is concentrated in specific bands,' he says. 'There are areas that are highly productive [supporting large amounts of life] where currents intersect, or where sea-ice exists, but the rest of the southern waters have very low biomass [mass of living things] compared with the sea-ice. Less ice means less area to grow.'

Ralph believes it is important for him, as an authority in the field, to spread the word about the dangers of global warming. He adds that his team in Antarctica is one of very few groups that have been able to estimate the productivity of the sea-ice. This work is being fed into global models of environmental change.

He says, 'I try to make sure the public is aware of our research. When I am meeting politicians and advisers I stress the importance of dealing with climate change, and of learning about these problems quickly.'

The Antarctic team works in October and November, arriving with the first transport ship of the summer and leaving as the sea-ice starts to break up. 'One year we were there in December when the ice began to melt and it started to get really dangerous,' Ralph says.

If contributions to such significant fields is not enough, Ralph is also conducting or supervising studies on microalgae in other locations. Much of his work now focuses on trying to unravel how microalgae use light. This leads in several directions. One path sees him investigating the possibility of improving algal growth efficiency in order to advance the prospects of turning farmed algae into biofuels.

Another path involves studying light use in nature as part of the Integrated Marine Observing System, a project Ralph says 'seeks to better understand coastal processes, developing a bio-optical capacity [sensors that use light to detect the quantities of living things] to understand how the algae use light so we can understand coastal productivity.' The microalgae act as the primary producers in the systems he is studying, feeding the fish population and supporting the entire ecosystem.

Climate change is already affecting the productivity of these microalgae, and Ralph is keen to understand how. 'The East Australian current is moving further south, making it warmer, but that is not always going to bring higher productivity. It's changing the amount of nutrients, changing productivity along the coast. We're trying to understand how changes in the physics alter the biology of the animals so we can feed this into the climate change models to understand what's going to happen.'

Ralph is also keen to find better ways to measure the carbon absorbed and removed by coral reefs, seagrass meadows and other ocean ecosystems so that changes in these can be fed into future carbon accounting. 'There's some technology developed for measurements on land, we're developing this for carbon fluxes in water.'

However, asked to name the research he is most proud of, Ralph says the team is 'making major inroads into identifying the biochemical trigger that causes coral bleaching.'

Reef restoration

The world looked in horror at the 2004 Boxing Day tsunami in which over a quarter of a million people around the Indian Ocean died. For most, the only way to help was with a donation. However, Dion Trevithick-Harney found that his scientific training enabled him to make a real contribution to Thailand's recovery, and in a more satisfying way.

At the time Trevithick-Harney was an Honours student in marine science at Curtin University of Technology. His thesis was on methods of restoring coral in damaged reefs, so when word went out on a coral email forum that reef construction group Reefball.org was seeking 30 volunteers to help restore a damaged artificial reef off Racha Island, he was at the front of the queue.

Applications were received from 370 people from around the world. The 30 selected volunteers paid their own airfare, but the local resort provided accommodation and food during their stay.

Racha Island, lying just south of Phuket, did not suffer as badly as many sites. Video footage shows someone being sucked out to sea in the waves, but everyone known to be on the island at the time was accounted for afterwards as survivors. Nevertheless, the resort was so badly damaged it officially shut down for more than six months.

The artificial reef Trevithick-Harney helped to restore was established because the local authorities had become aware of how badly scuba-diving tourists were damaging the area's precious natural reefs. An artificial reef closer to shore provided a place where people could learn to dive and watch fish; once they'd learnt to avoid damaging the precious corals they were allowed to visit natural reefs.

With many natural coral reefs damaged by the tsunami it was essential for the artificial reef to be restored in time for the tourist revival if the local ecosystem was not to suffer further damage.

Trevithick-Harney arrived in April, and found that damaged coral could be restored 'quite easily. It had been broken into pieces ranging from several tonnes down to lumps the size of a fist. We moved the smaller pieces into a nursery, and made sure they were facing upright and getting plenty of direct sunlight.

'Once the coral was stabilised we went down with pliers and cut the coral into 5–10 cm pieces, took them across to a table and cemented them into a plug, which was then dropped back into place,' he said. The artificial reef is about half a metre high and filled with holes that the plugs could be cemented into.

Large parts of the reef had been turned upside down in the tsunami, with the coral dying before Trevithick-Harney and the team could get to it, but substantial sections had survived relatively intact. Locals reported there had been no fish in the bay for a month after the disaster, but some returned prior to the restoration effort.

All members of the recovery team were involved in the hands-on restoration, but Trevithick-Harney and the other two marine scientists were given extra responsibilities.

They also spent a lot of time teaching the other team members about the coral species they were working on. On land, the area's recovery has been mixed. Many seaside trees were devastated, and one species that serves as the only host for a particular species of caterpillar has been wiped from the island completely.

Trevithick-Harney says he was inspired to take up marine science at an early age, having seen a Jacques Cousteau documentary as a child. 'I never got it out of my system,' he says. Further documentaries maintained his enthusiasm.

Trevithick-Harney's Honours thesis is on propagating *Favites* brain coral. These relatively streamlined corals don't fragment naturally, so unlike many other corals they rely on sexual reproduction. 'There's evidence that copper, tin and oil inhibit their reproductive capacity,' Trevithick-Harney says. Consequently *Favites* struggle to recolonise damaged reefs in polluted environments.

Trevithick-Harney tested the survival prospects of small pieces of coral he broke off under 'starvation conditions' – that is, sunlight but no food supply. While *Favites* is not used to fragmenting naturally, Trevithick-Harney found it recolonised well and when multiple polyps were seeded he achieved a very high survival rate. Even with single polyps some species of *Favites* returned very high success rates, demonstrating that this important type of coral can be regrown after damage.

Trevithick-Harney's recovery mission came as he was about to apply for a PhD scholarship in marine biology, but had yet to settle on a topic.

After finishing his Honours thesis Trevithick-Harney returned to the industrial sector working in a variety of roles, ranging from a chemist for an iron ore mine to now working at a power station as a chemical and environmental specialist. 'My passion for marine biology continues and has led to managing an aquaculture farm in Puerto Rico raising clown fish and other ornamental marine fish for the aquarium market in the United States,' he says. 'As for the future? I see it as a mixture of following my passion for the coral reef and working in the industrial sector.'

Eight eyes, no brain

How can a creature without a brain swim long distances against the tide and wind? How does it even decide where to go? Those are the questions Matthew Gordon is trying to answer. His research requires him to catch some of the world's most venomous species 'by hand' and stick electronic tracking devices to them.

Gordon is doing a PhD in marine biology at the Tropical Australia Stinger Research Unit at James Cook University. Although he took a number of years off to travel the world before starting his doctorate, it was a fairly inevitable path for him. Growing up he says he 'was always studying science, and particularly the biological sciences'.

Gordon says box jellyfish have fascinated him for some time, and they were the subjects of his Honours thesis. He is now working on the largest and most dangerous species, *Chironex fleckeri*. Jellyfish are hard creatures to study, and not a lot is known about them considering how common they are. However, with major blooms breaking out in areas where competing fish have been overharvested it's becoming increasingly important to understand their biology and behaviour.

True jellyfish cannot travel independently of wind and current, but box jellyfish can. Their tentacles are not used for 'swimming', although they are contracted for better water resistance. Instead they use what Gordon calls 'a sort of underwater jet propulsion'. The jellyfish's bell contracts to force water out, and a ring of tissue called the velarium narrows the opening so the water squirts out under high pressure.

Gordon's studies involve sticking ultrasonic tags onto the jellyfish so their motions can be observed. He is not the first to do this, but he says 'previous tags were too heavy so they interfered with the normal behaviour of the jellyfish. These tags weigh about a gram,' which is similar to some of the small fish *C. fleckeri* prey on.

The tags are attached using a glue borrowed from tissue surgery and stay on for about three weeks. In order to do the tagging, a team of two go out in a boat, with one steering while the other stands at the front looking out for jellyfish. When one is spotted the individual at the front has to jump in the water and catch it. Full-body wetsuits are used to avoid being stung.

The results of the tracking have been remarkable. One jellyfish was observed to travel 7 km in 17 hours without help from wind and current. Another managed 10 km over 20 hours, and Gordon says the last part of the journey was into an estuary against the outgoing tide. They can move considerably faster over short distances, presumably to chase prey or avoid predators.

Gordon notes that while the box jellyfish tentacles can stretch for metres, the bells themselves are not particularly large, making the journeys more impressive. By tracking a sexually mature adult in its journey up an estuary, Gordon was able to confirm the theory that box jellyfish enter them to reproduce.

It is hoped that an understanding of what determines jellyfish movement will make it possible to reduce the number of stingings as people come to understand how to stay out of their way. In Australia, the past century has seen about one death from jellyfish stings per year, but in South-East Asia the rate may be 100 times as high. Already Gordon's studies have shown that jellyfish take a siesta in the middle of the day, resting on the bottom of the ocean with their tentacles spread out like a relaxing cat.

Gordon confirms that box jellyfish are more likely to try to avoid humans than to deliberately attack them. What remains a mystery is how box jellyfish decide where they are going. They don't have brains or even a central nervous system. Nevertheless they will move away from shadows.

Box jellyfish have 24 eyes. Sixteen of these are light-sensitive pits incapable of forming an image; the other eight are more complex with sophisticated lenses, although the placement of the retina suggests that the resulting images would be blurred. Gordon says that a Scandinavian team collaborating with his group is trying to unravel the mystery of the jellyfish eyes, including how the animals are able to make use of the information their eyes provide, possibly through some sort of interpretation at the site of each eye.

Mathematicians

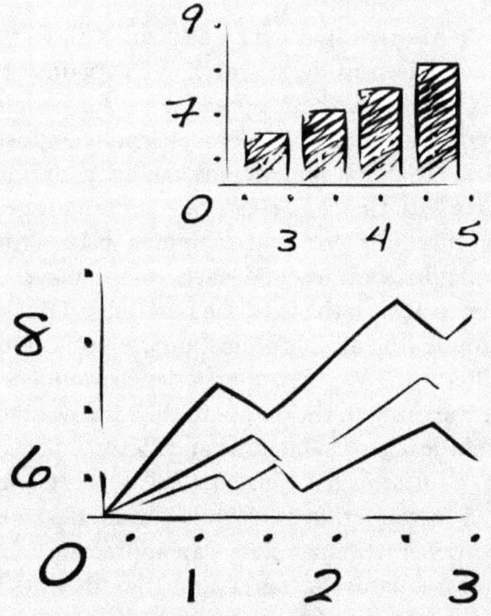

Mathematical art

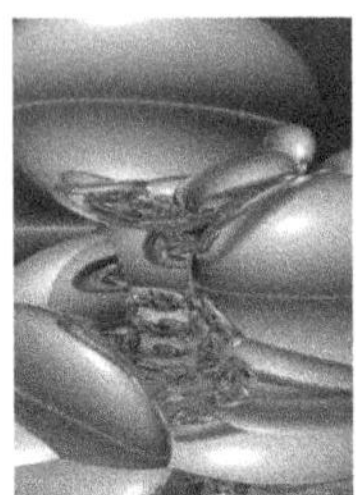

Dr Cameron Jones is a mathematician, an artist and a nightclub owner and manager. The separation between the artistic and scientific worlds can be wide, but Jones is trying to bridge the gap.

'All my digital art works are a reflection of things that have arisen in my research,' he explains. 'I am interested in art for the aesthetic appeal it has for people, but I also hope people are able to reflect on it and recognise the tremendous value that mathematicians can bring to the arts.'

'My artwork plays with time and space by manipulating simple shapes like spheres and cubes in three-dimensional computer space. Things look very different at different magnifications, and each image plays on this theme to encourage the viewer to absorb the extreme relativity of physical reality.'

'Mathematics, modelling and visualisation are not too different from storytelling, film and literature. We seek to experience numbers as a picture. This is how pictures hide meaning in the detail, in the same way some formal equation explains different levels of behaviour at different stages of the proof. For example, my work *The Cantor Set Contains It All* uses Cantor sets (a type of infinite set used as an introduction to fractals[1]) and reflects on space-time that mathematicians and non-mathematicians can appreciate.'

Jones sees his art as a natural progression from his work in image analysis and his interest in 'the power of technology to increase resolution and our sensing ability.' This took him into photographing chemical shapes, and he describes working on creating art from mathematical modelling as a 'natural transition'.

While bringing art and mathematics together is a challenge, uniting both with a nightclub in trendy Smith Street, Collingwood, is more unusual. Yet Jones co-owns and co-runs the Blue Velvet Bar & Nightclub, a venue known for its poetry nights, goth and industrial music, and edgy theme nights. Estab-

1 Fractals are mathematical shapes that repeat themselves on different scales, so that the more one focuses in the more one sees the same patterns repeated. They have proved extremely significant for mathematical applications from predicting weather to improving transmission speeds of videos.

lished over a decade ago, the bar has expanded and become one of the survivors in a rapidly changing industry.

Jones uses the venue as an artistic space as well as a testing ground for studies analysing what music is likely to prove successful. Jones doubts, however, that mathematicians will ever replace disc jockeys, let alone composers. 'DJs are fairly intuitive ... I am doubtful that fractal music – turning equations directly into music – will work. I don't think you can do away with people.' Instead he hopes his work will help record companies find ways to release and distribute music with maximum appeal.

DJing one night Jones discovered that yeast on the surface of a CD can change the music. He's since experimented with remixing tracks by growing bacteria and fungi on the CDs he plays.

All this might be regarded as a distraction from the serious work of mathematics. However, while Chancellery Research Fellow in the Centre for Mathematical Modelling at Swinburne University Jones found time to produce numerous papers on the application of maths to commerce, engineering and biotechnology.

Jones left Swinburne, not because his art was too distracting, but because the university tried to force him to take too heavy a teaching load. He now works as an independent scientific consultant, as well as having more time to run Blue Velvet.

Jones's Bachelor of Science degree at La Trobe University included a major in microbiology, and before specialising in mathematics he spent eight years in microbiology labs, with an emphasis on the study of fungi.

He doesn't see the artwork as simply a hobby. 'I feel that, as a scientist, one has a duty to inform and explain. However, mathematics is inaccessible to most people when it remains in its purest form, so this is the most perfect avenue for bringing maths to the masses.'

The beer scheduler

Slim Dusty may have sung about a pub with no beer, but that was before mathematical modelling became an integral part of a brewery's operations. As Victorian regional manager of operations planning with Carlton and United Breweries (CUB) for six years, Dan Boulton's job was to keep much of Australia supplied with its favourite brands of amber fliud.

Boulton's mathematical modelling ensured that the right amount of beer, in the right sort of packages, went to all the places that would need it. Interviewed one January, he sounded tired from a truly festive season when a heatwave combined with the closure of CUB's Sydney brewery to rocket demand for the products he managed.

Since his brewery produces 4.6 million hectolitres of beer per year – double Australia's next largest – there was plenty to do at any time. It's barely a decade since stocks sometimes ran out over Christmas.

According to Boulton, most of the people filling similar roles do not share his background as a maths major (at the University of Melbourne). 'They come from diverse backgrounds – lots of maths, but also engineering, economics, computer science, and some people without degrees who worked their way in through the business.'

Nevertheless, Boulton found his studies useful. 'We use very sophisticated scheduling software which uses the algorithms[1] I learnt at uni. I'm not getting my pen and paper out myself, but you need to know what the tool is doing to use it, to understand the ways inputting the data influences the outcome.

Beers are not all the same for scheduling purposes. 'Some products take longer to ferment, yeasts take different times to grow. There are constraints on packaging, needing to send a constant number of trucks,' Boulton explains. He previously worked at CUB's head office as part of a team that received the demand projections from marketing and provided instructions to CUB's breweries on how much product each needed to supply. His transfer resulted from a crisis at the Abbotsford brewery.

'One guy went on long service leave, one resigned to take another job, and one had a heart attack, so you had one person doing the job of four. They asked

1 An algorithm is a step-by-step procedure for solving a problem, for example a series of calculations in maths.

me to step in. It was pretty hectic, but when the worker on leave came back I wrote a report on how things could be improved.'

That report – suggesting that brewing, packaging and transportation be brought together rather than managed separately – impressed his bosses and Boulton was offered the opportunity to take charge. The position suited him, with less time in front of a computer and more in the brewery learning about 'things like safety, which is our number one priority, and quality, which is number two. Like the reuse of materials – everything here is reused. Spent grain goes for cattlefeed, yeast for vegemite or fishmeal. Brewing produces a lot of CO_2, which is then used in the bottling stage.'

Boulton learnt a bit about other sides to the job as well. 'Sometimes there's a more efficient way to do something, but someone prefers to do it their way,' so he needed to accommodate.

While admitting that he witnessed some extraordinary foul-ups in his job and even factors these into the safety margin for inventories, Boulton is too discrete to reveal any specifics. Instead, when asked to nominate a particular incident from his career he talks about the excitement of working on the distribution of beers to each event at the Sydney Olympics.

Despite a fondness for the product, Boulton didn't plan on working at a brewery while at university. 'I wasn't someone who went to uni to get a job. If I was, I probably would not have done maths. I did [maths] because I liked it at school and was good at it. I was also interested in physics and computer science, but had a general preference for maths, and that's the way it stayed.'

Despite the lack of vocational planning in his studies, his skills have proved very much in demand. This is a trend Boulton expects to continue. 'Most companies used to just make things. But as the price of production falls, and the price of transportation and storage rises, there is huge scope for efficiency savings in inventory distribution.'

The challenge is particularly important for beer manufacturers. In 1980 the average Australian drank 130 litres of beer per year. Now it's only 70 litres. 'We got hit by a range of factors together. The baby boomers who were big beer drinkers reaching an age where they couldn't drink as much, the changing ethnic mix, rising taxes, a shift to wine and mixed drinks.'

Nevertheless Australians are not likely to give up drinking beer any time soon, which made Boulton happy since the job came with two free slabs per month, extra during the holidays, and a free bar every night.

Despite these incentives, after too long a job can become repetitive. Boulton moved on. He is now employed by Coles Supermarkets where he works on the supply chain and replenishment process. 'Don't you hate it when that one ingredient needed for your recipe is out of stock at the store?' Boulton asks. 'How does a supermarket chain ensure customer satisfaction across so many products and so many stores – and how can it do it cheaply and efficiently?

'What is the best way to get $20 billion dollars worth of groceries to 750 stores throughout Australia from over 2000 suppliers based in Australia or abroad? Now take into account some further complexities of different product types, such as fresh produce, meat, dairy and other products with very short shelf life, general merchandise sourced from overseas with long lead times, seasonal products (Easter eggs, Christmas), promotional discounts driving volume increases, competitor impacts, distribution centre and transport capacity constraints, and demand variation with the weather (soup in winter, soft drink when it is hot) and you have a very big puzzle to solve!'

How effectively Coles meets customer demand while minimising the cost of the supply chain (over $1 billion annually) is a key to business success. Problems of this size and complexity can only be solved with mathematical models and computer processing power. It's a tough problem to solve, and one that constantly redefines itself – but it never gets boring. 'A background in mathematics gives you the tools to formulate a plan across this vast network of suppliers, distribution centres and stores to ensure that when the special recipe calls for the vital ingredient, it is always there.'

Statistics brought to life

People who like science may be surprised to find a statistician in this book. Archaeologists, palaeontologists and even occasionally mathematicians get to star in Hollywood blockbusters, but the popular view suggests statistics are tedious and statisticians highly tolerant of boredom.

Rob Goudey sees things differently. 'I get a buzz from applying statistical methods and seeing models just "fall out" from the data for the first time,' he says.

Goudey is aware of the misuses to which statistics can be put. The signature on his emails includes a quote from author and historian Andrew Lang: 'He uses statistics as a drunken man uses lamp-posts: for support rather than illumination'. But the choice also indicates his passion for statistics used properly, and the pride he takes in potential outcomes from his research.

As statistician for the Victorian Environment Protection Authority (EPA), Goudey analyses environmental problems such as water quality in Port Phillip Bay and smog levels over Melbourne. His skills are applied to a broad range of scientific disciplines, and the impacts on others can be swift.

Goudey's interest in using maths in science was sparked when a Year 11 biology teacher gave each student seeds of native trees and asked the students to grow and study the plants throughout the year. He came across an article describing how the need to withstand bending stress influenced the shape of stems and branches. This led him to try to model various aspects of plant shape, a calculation that failed because 'I only had high school maths'.

Goudey defies the stereotype of the statistician who prefers numbers to living things. After high school he worked for a while for the Victorian Forestry Commission. 'I hated the idea of lopping trees and so refused to use the chain saw,' he recalls. 'On one of our trips a tree that the other technician was trying to finish off with the axe twisted as it fell, landing on top of the chain saw and crushing it. A fitting end to our trip.'

Forestry was followed by four years as a draftsman before Goudey concluded he really wanted to study science, which he did in the form of a Bachelor of Education majoring in mathematics and biology at Rusden (now part of Deakin University).

'In contrast to many of the other students I found biology much harder than maths,' Goudey says. 'In maths the fundamental concepts were taught to us and we just had to use them to solve problems. It was usually made obvious which concepts applied to which problems.

'But biology was more complicated than this. The fundamental concepts were often not clear-cut, and it was often not obvious which concepts applied to a given problem and in what way.' After a stint at the Royal Children's Hospital analysing data on children with anorexia, Goudey moved to the EPA, where he's been for 21 years.

He's currently involved in a collaborative project with CSIRO and the Victorian Department of Sustainability and Environment to develop statistical analysis and reporting tools as part of the Commonwealth Government's Australian Water Data Infrastructure project. Goudey is providing graphical and statistical methods that will help water scientists detect unexpected changes in water quality by comparing measured water quality to values predicted using models. Departures from predictions may mean that the models need updating or that the environment is changing in a way we didn't expect. He says, 'I hope they will help in detecting early warning signs of effects that may be due to climate change.'

Goudey counters those who see computer modelling as abstract and unrelated to people's lives by describing the outcomes from his work on deriving water quality objectives for Victoria's streams. 'The old objectives were based on human beneficial uses, but they failed to take into account ecology and natural background variations. The new objectives take into account the natural variations and scientists' opinions of what is achievable. Those objectives are now law in Victoria.'

Goudey recognises that good science and good environmental decision-making depend critically on good quality environmental data, and he works with scientists to develop data quality management plans for their datasets. Quality data is a part of any scientific study that's often merely assumed, and so it's sometimes overlooked, but the impacts of poor quality data can be devastating.

Sharing expertise is a major part of his job. Consulting with biologists, meteorologists, chemists and toxicologists, he enjoys being able to delve into many other areas of science while offering an alternative perspective to the people he works with. The learning goes both ways. As the statistician John Tukey said, 'The best thing about being a statistician is that you get to play in everyone's back yard.'

Medical scientists

A model scientist

For most scientists, modelling is done with mathematics or a computer. For Dr Valerio Vittone, however, there are other ways.

Vittone is studying the molecular structure of herpes simplex virus (HSV) type 1 as part of a PhD in molecular biology. But he's also had a successful career as a model, and several as a businessman.

Vittone was born in Italy, and explains: 'I started modelling when I was 18, when I was scouted in Milan, and after a year I did an editorial for *Vogue Italy*.' Elite, a top Paris agency, signed him up and flew him around the world for assignments.

Vittone now models 'mostly for charity or publicity', for example modelling clothes for the fundraising campaign Designers Against Aids (part of Designers and Models Against AIDS). Instead he opted to study science at the University of New South Wales, having made Sydney his home at the age of 22. Despite receiving an offer to work in the United States, Vittone says he decided he would prefer to do at least his PhD in 'a country I feel closer to,' accepting an offer from the University of Sydney.

Vittone is appalled by the low status and funding of science, particularly for young researchers. He says that one reason he has stuck with research is that he hopes 'to create a momentum, a force of young people who would be successful in science,' rather than leaving for the financial sector. However, despite 'not wanting to become a businessman,' he has founded five companies with interests as diverse as importing high-tech construction materials and radiation-free earpieces for mobile phones. Vittone-designed women's shoes have also achieved success. Most recently he's taken to marketing his mother's hand-made pasta.

Combining his three fields, Vittone created the Renaissance Ball, to help fund equipment and researchers in need at the university's Westmead Millennium Institute. Designers, models and celebrities donated their time and skills to raise $350 000. 'I'm not truly a businessman,' Vittone says. 'I'm a scientist, but I see how my colleagues suffer from a lack of funding, so I want to try to do something about it.'

'I was inspired by the Renaissance because it was a period in Italy when beauty and science were together in harmony,' Vittone said. He hopes that this Renaissance really will be a new beginning for science.

Vittone's work has made him an ambassador for Designers and Models Against AIDS, but he hopes the Ball also increased awareness of the plight of young scientists, something Vittone is particularly concerned about. He appeals for senior scientists to treat their young researchers with respect, allowing them to experience 'the mystery of science' rather than being treated 'as employees and producers of papers'.

While admitting that 'sometimes it is hard to spend all day in the lab' with so many more glamorous jobs available, Vittone plans to keep to his current course. He finds it sad that other scientists often think he is mad, saying 'This should be the glamorous job.' He considers the 'lack of a relationship between the benefit an individual can bring [through science] to the financial benefit obtained tragic'. Through his fundraisers Vittone seems himself as 'something of a Robin Hood,' redistributing wealth to the more deserving.

Given all this, it's perhaps not surprising that *Cleo* nominated him for Bachelor of the Year 2008. Vittone hoped the profile would assist his fundraising efforts. He didn't win, despite 120 000 votes and a 2700 strong Facebook site supporting the nomination. Commentators speculated the judges might have been put off by his intelligence and altruism.

Vittone's own science work has plenty of potential for human benefit. HSV-1 (responsible for cold sores rather than the genital herpes of type 2) is extremely common – up to 70% of the population carries it, although relatively few show symptoms. The effects, however, can be as serious as blindness and brain inflammations. Some research has linked it to Alzheimer's disease.

'My research has led our team to find key protein–protein interactions on the surface of HSV,' Vittone says. 'The information provided by these key interactions will be used to make peptide analogue drugs, short amino acids that could selectively "stick" to the surface of the virus, stopping infection by preventing it from travelling or/and self-assembling.'

These 'magic bullets' could be five to 10 years away. While a cure for herpes will make many people happy, there's a bigger payoff. HIV infections are much more likely to occur in people already infected with HSV, so Vittone's work could reduce the infectivity of HIV.

HSV uses the micro-tubular structure of the nervous system, penetrating eventually to the masses of nerves called ganglia in the spinal cord. Its ability to reach so deeply into the body makes it an ideal vector (carrier) for gene therapy, in which a healthy gene is attached to a vector and carried to the part of the body needing repair. However, Vittone believes this may not be an unqualified good, saying: 'The public needs to know the impact what we are doing in the lab will have on their lives,' and debate whether they want some of the more controversial aspects, such as the capacity to 'treat' people for things that are normal, but sometimes considered undesirable.

Skin for life

The ill wind of 202 deaths in the 2002 Bali bombing blew with it at least one piece of good. The heroic work done by Professor Fiona Wood of the Royal Perth Hospital significantly raised the profile of her discoveries, and assisted with the fundraising required to enable her to concentrate on further research that may save the lives of many future burns victims.

Wood is the inventor of spray-on skin cells, a technology that substantially contributed to the treatment of many of the burns patients from Bali, and has since been used in thousands of other cases. This technology, and the profile it gained after the Bali bombing, led to her being awarded 2005 Australian of the Year.

Where previous techniques of skin culturing required 21 days to produce enough cells to cover major burns, Wood has reduced that period to five days. This is important because, as Wood says, 'The longer you wait the more your chance of not surviving.' Furthermore, Wood found that scarring is greatly reduced if replacement skin could be provided within 10 days.

Wood's research into improved methods of cultivating replacement skin cells began in 1993, and the spray-on skin cells were first available in 1995. In order to reduce by three-quarters the delay before skin could be transplanted, Wood had to do far more than replace sheets of skin with sprays. Secrecy agreements prevent her revealing most of the developments involved in the progress, other than a vague reference to 'paying attention to a lot of details'.

Marie Stoner, the scientific director of Wood's company Clinical Cell Culture, is slightly more forthcoming. In an interview with the magazine *Medical Forum* she relates the struggle to get skin cells to grow faster, saying, 'We even thought of trying magic or talking to them.' The pair was spared such a venture beyond the realms of science through the discovery that grafted cells grew more quickly on the patient than in the lab, allowing them to spray the cells onto the burns with much shorter delays.

Clinical Cell Culture (now Avita Medical) markets ReCell®, a stand-alone, rapid, autologous[1] cell harvesting, processing and delivery technology. ReCell®

---

1 Autologous transplantation is from one part of a person's body to another part of the same person.

has been designed for use in a wide variety of wound, plastic, reconstructive, burn and cosmetic procedures. It enables surgeons and clinicians to treat skin defects using the patient's own cells in a regenerative process, accelerating healing, minimising scar formation, eliminating tissue rejection, and reintroducing pigmentation to the skin.

The procedure is performed on site utilising a patented and proprietary 'spray-on' technique. It takes approximately 30 minutes to complete and does not require laboratory facilities.

Medical research was not a lifelong ambition for Wood. Indeed she says, 'I came from a Yorkshire coal mining village ... I hadn't thought of it until it was clear I had the marks to do it. I thought it sounded like a nice idea.'

Once she began studying medicine at St Thomas's Hospital Medical School in London 'It became clear to me I would be a surgeon'. She chose plastic surgery because, 'It's creative, about rebuilding not removing. It's also innovative, lots of opportunities for new technology.'

The inspiration to work on burns occurred in East Grinstead, where the New Zealand surgeon Sir Archibald McIndoe did pioneering work during the Second World War. Sir Archibald died several years before Wood came to his hospital, but the centre for excellence he had established convinced her 'we need to maintain the improvement'. Shortly after the Bali bombing, Wood found a quote from McIndoe with considerable poignancy for the modern age: 'The privilege of dying for one's country does not equal the privilege of living for it.'

After coming to Australia in 1987 Wood became head of the burns unit at the Royal Perth Hospital. She has treated more than 1500 burns patients, along with many others suffering from soft tissue loss. She has also found time to have six children, leading to a working day which would make most people tired just reading the timetable. It's not hard to believe that as a child she aspired to a career as a sprinter.

Wood can't explain where her energy comes from, simply saying 'I've always been a bit manic ... always done things quicker. It took me a long time to realise I was not the norm. I intuitively move quicker and hate to waste time.' It was a characteristic that stood her in good stead when the Bali tragedy occurred.

Working 18-hour shifts for a week, Wood treated more than 30 patients, many with very severe and extensive injuries. While the team had done disaster planning based on bushfires or an explosion on the oil platforms of the North-West Shelf, they had experienced nothing like this. In a typical week the hospital would admit around three new burns patients, and some of the wounds resulting from the explosion were different from those found in even serious ordinary burns victims. Wood was uncomplaining, however, noting that she was 'getting her usual five hours sleep a night'.

The higher profile has allowed her to contract a 'five-year fundraising plan to 12–18 months' in her efforts to give up private surgery so she can concentrate full time on public work. 'Nevertheless,' she says, 'we need to keep the ball rolling' if she is to be able to maintain that focus.

Wood's goal is 'the holy grail, to regenerate composite skin, not just the epidermis [the outermost layer of skin].' While such an achievement is many years off, she believes a more short-term achievement will be 'the use of artificial intelligence in assessment, so you can work out what you are doing – how deep is the burn, how thick the scar.'

A parasitologist and clergyman with 'a passion for poo'

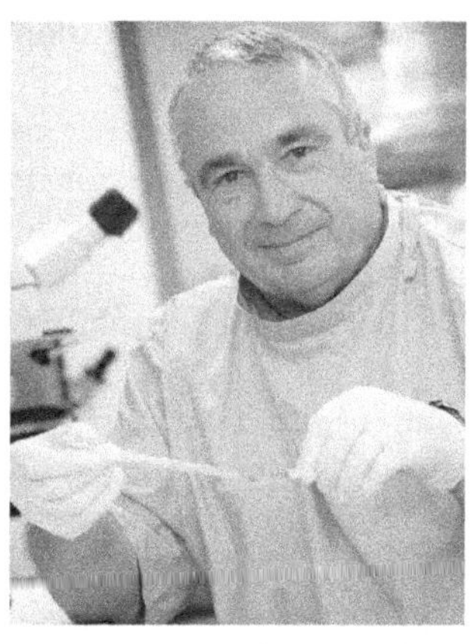

Associate Professor Reverend Dr Wayne Melrose sees his job as 'healing the sick'. He diagnoses and treats parasites capable of destroying the lives of millions on weekdays, and ministers to a congregation on Sundays.

He 'makes no distinction between the two' professions. However, since his weekday work involves identifying parasites in faeces he 'assures my congregation that I always wash my hands carefully between my Friday and Sunday work'.

'I have a passion for poo, blood and parasites,' Melrose says. 'Some people might be put off by a collection of blood and faecal samples, but to me they are a treasure trove of exciting possibilities. I just can't get over the "wow factor" when I see something new to my experience, or read about the discoveries others are making, and I see that enthusiasm rubs off onto my students and colleagues. As a result, the conversations in our departmental lunch room are not for the faint hearted – they are often about the latest parasites we have found, and inevitably turn to the topic of "poo".'

Melrose was inspired towards science by an excellent teacher at high school, and in the 1960s started work as a medical laboratory scientist in New Zealand. He moved to Melbourne in 1974 and completed a Bachelor of Applied Science at the Royal Melbourne Institute of Technology in 1999. In the early years he concentrated mainly on the biology and biochemistry of blood, including hereditary blood disorders, the effects of alcohol abuse, and the blood changes associated with sudden infant death syndrome. The latter research gave him the opportunity to go to Antarctica. One of the theories about SIDS was that babies forget to breathe, and the resulting lack of oxygen puts an enormous strain on metabolic systems. 'Elephant seals can hold their breath for a very long time and we came up with the crazy idea that we might get an insight into SIDS by studying how seals cope with lack of oxygen. We did not find a link between seals and SIDS but we did publish the first-ever paper on erythrocyte [red blood cell] metabolism in seals,' Melrose says. 'This led on to other work on penguins, various marsupials, and even lizards! You never knew what you were going to find in that sack under the laboratory bench!'

Although his main work was on blood disorders, he had a long-time interest in parasitology and tropical medicine that was further fuelled by working in a hospital in Papua New Guinea in 1985. In 1992, he had a 'mid-life crisis' and moved to Townsville to begin a Masters degree in public health and tropical medicine at James Cook University (JCU). This led to doctoral studies and in 2002 he was the first person at JCU to be awarded a Doctorate of Public Health.

In 2004 he was appointed as director of the World Health Organization (WHO) Collaborating Centre for Control of Lymphatic Filariasis and Soil-transmitted Helminths (tropical parasitic diseases). He holds a number of advisory posts within WHO and was appointed as an associate professor in the School of Public, Tropical Medicine, and Rehabilitation Sciences at JCU in 2006. He is a Fellow of the Royal Society of Tropical Medicine and Hygiene, a Fellow of the Australasian College of Tropical Medicine, and a Fellow of the Australian Institute of Medical Scientists.

Melrose says he has been a 'practising Christian all my life', but the idea of becoming a priest came later. He trained to be a medical missionary and, while he never reached the missionary fields, the trend to ordained clergy working in other jobs has allowed him to combine the two pursuits.

'Four billion people are infected with intestinal or blood parasites worldwide, many of them children living in poverty,' Melrose says. 'They cause widespread misery, but because they seldom directly cause deaths they are often over looked. There are effective cures for almost all the parasitic diseases. The problem is finding out where they are and getting the treatments to people.'

Melrose has been particularly involved in fighting lymphatic filariasis, the cause of elephantiasis. 'It's a dreadful disease – enlargement of the limbs, scrotums down around the knees. It also attacks the kidneys and reduces the efficacy of the immune system.'

Melrose says the devastating impact of elephantiasis was first brought home to him about 15 years ago when he met a girl in Papua New Guinea with the first symptoms of the disease. She asked him: 'What man will marry me now I have this disease?' At the time it was believed that nothing could be done for patients who had developed elephantiasis, but new research has shown that a combination of increased hygiene, treatment of skin infections and gentle exercise can prevent the progression of early disease and markedly improve the quality of life of those with more advanced disease.

The World Health Organization has set a target to eliminate lymphatic filariasis as a public health problem by 2020. The strategy is a simple one

– treat the total population at risk with a combination of two common drugs for a minimum of five years – but it requires political will at all levels of government, and a lot of funds. The campaign against lymphatic filariasis is actively supported by the drug manufacturers GlaxoSmithKline and Merck, who donate their products free of charge, and by organisations such as the Bill and Melinda Gates Foundation and academic institutions. The campaign is going very well, with most of the Pacific Island countries approaching the point where they can be declared filariasis free. However there are still big challenges ahead: Papua New Guinea, which has the highest prevalence of filariasis in the world, has yet to start a national campaign.

Intestinal parasites, Melrose's other enemy, are often 'treated as a fact of life in the tropics,' he says, even though they 'stunt growth, cause anaemia, and can lead to learning difficulties. If you have a child who is always itchy, hungry and whining they don't learn well. You need a multi-pronged program to control them. Awareness of hygiene such as washing hands, toilets that work, clean water and drugs for treatment – they all need to come together or you just go round in circles.'

Parasitic infections can cause people to be more susceptible to major killers such a HIV/AIDS, tuberculosis and malaria. Global warming is also presenting a challenge to parasitologists. We may see organisms from warmer climates causing illness in more temperate areas, including Australia.

'I have never regretted my choice of science as a career, and I really don't see myself ever retiring completely, I am having too much fun!' Melrose says. 'I currently have projects running in Papua New Guinea, the Solomon Islands, the central Pacific and Timor-Leste, and have worked in, or visited, over thirty countries in the course of my work.'

While some scientists, parishioners and clergy 'raise their eyebrows' at Melrose's combination of roles, he reminds them that he is following a long tradition of Christians in science and public health. He tries to avoid those areas where science and the church sometimes clash, saying 'I try to stick to things Jesus did, like bring hope and healing to people in need. I don't think Jesus was very interested in arguing creation versus science, or about evolution. We often focus on the divisions and ignore the fact that scientists and theologians have a lot of common ground.'

At the viral frontline

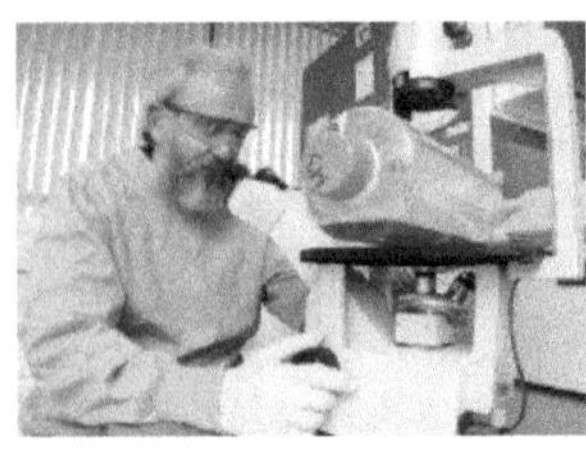

Many medical scientists spend their working lives battling a single disease. Dr David Boyle, on the other hand, has been part of the struggle against many diseases, both human and animal, including roles fighting some of the highest profile viral outbreaks of recent years.

Boyle is a Senior Principal Research Scientist at CSIRO's Australian Animal Health Laboratory (AAHL). His primary role is in the creation of diagnostic tests and vaccines for animal disease. However, as avian and swine flu (2004 and 2009 respectively) have reminded us, diseases that infect animals can jump the species barrier to threaten humans.

In 1982 Boyle joined the AAHL and investigated the prospect of turning viruses from threats into guardians, specialising in the fowlpox virus. 'The primary focus of the fowlpox research was to protect poultry,' Boyle says. 'Human applications were a spin-off.' But what a spin-off. Fowlpox viruses offer prospects for better delivery of vaccines because they can target cells affected by or vulnerable to disease, taking the vaccine's material with them. While many potential viruses could be used as vectors,[1] Boyle and his colleagues have demonstrated fowlpox's value: it is large enough to carry extra genetic material and is safe for humans, but it is still capable of making human cells produce copies of the material it carries.

Australian biotech company Virax Holdings is trialling fowlpox virus to carry gamma-interferon (a promoter of cellular immunity) and genes from the HIV virus to the cytoplasm of cells [the part of the cell enclosed in the plasma membrane, not including the nucleus] of people with HIV. Once in the cytoplasm, the fowlpox virus produces antigens (molecules the immune system recognises) and more gamma-interferon. It is hoped that this technique will boost production of T-cells, the cells attacked by HIV, and match the ability of anti-retroviral drugs to keep HIV under control without the serious side-effects of these drugs.

1 In molecular biology a vector is a virus or other substance used to transfer genetic material into a cell.

This could greatly improve the life expectancy and quality of life for those with AIDS. However, only a preventative vaccine will finally solve this tragedy. Here Boyle also has a role to play. Animal trials at the Australian National University's John Curtin School of Medical Research have shown potential for the 'prime boost' strategy, in which a DNA vaccine boosts the activity of T-cells before a vaccine carrying HIV antigens is administered.

The fowlpox virus has been chosen as the best way to deliver the boost phase. Human trials are still in the early stages, but Boyle says they confirm that at least 'there is no safety issue with fowlpox'. Boyle adds that use of the double vaccine in animals 'provides much higher levels of immunity than either on their own'.

Boyle believes fowlpox will prove valuable as a vector for vaccines against many different diseases, but notes that 'each vector has different biological characteristics that may make them more attractive for some diseases and less for others'. Consequently there is still a need to seek out other appropriate viruses.

Boyle has also confronted the Hendra virus (a virus spread by bats that caused two high-profile deaths in 1994, as well as killing numerous horses), and more recently avian flu. The AAHL has produced tests to enable quick identification of the virulent H5N1 strain that has decimated poultry in Asia, and led to over 100 deaths.

Australia has had outbreaks of a different form of avian flu on five occasions in the past, but Boyle says these were all 'small and contained' and spread by indigenous ducks, which were not themselves affected by the disease. 'The unusual aspect of the current variety is that it does kill ducks,' he says. Since a sick duck is unlikely to fly this far, Boyle says 'it's more likely to come into Australia on someone's shoes'.

'Over the last four to six years we've played an increased role in early detection, with recognition of the importance of nucleic acid-based detection systems,' Boyle says. Work has been done on 'avian flu, foot-and-mouth … anything we need to be ready for but is not here yet'.

Boyle says the AAHL team 'responds as the need arises, but have a priority list of diseases' to work on in the meantime. Its major focus is livestock disease as 'trade in livestock is 12–14% of Australia's external earnings'. Moreover, Boyle says 'experience in the UK is that livestock diseases not only impact on trade but local and international tourism, which is another 11% of Australia's export income.' Indeed SARS[2] and avian flu have proved that diseases do not need to reach our shores to affect tourism here.

2 The Severe Acute Respiratory Syndrome (SARS) outbreak in 2003 caused more than 700 deaths and widespread panic before it was brought under control. Since the worst affected countries were popular stopover points for tourists coming to Australia, many people cancelled trips here.

Boyle says he was 'always interested in science' but he never really developed a preference for a particular area: 'It could have been physics just as well.' However, he majored in microbiology for his undergraduate and Masters degrees at the University of Queensland. A scholarship from the Department of Primary Industries settled his career path. He spent three years in London studying rodent malaria as a 'model for understanding the [malaria-causing] parasite *Plasmodium* in humans'; on return he did a PhD in virology and immunology at the John Curtin School of Medical Research at the Australian National University.

If useful work was a priority, Boyle certainly took the right path. Not many scientists are simultaneously involved in promising trials against the world's leading infectious diseases while protecting one-quarter of Australia's export economy.

Flu busting cold case

Explaining the deaths of 50 million people would be the ultimate murder mystery, except that the cause was a virus rather than deliberate human action. Nevertheless, Professor John Mathews' work is a little like the 'cold case' detectives who use modern science to explain deaths that went unsolved when they occurred.

In his case the question is how the Spanish flu of 1918–19 managed to wipe out so many people so quickly. It was inevitable that Mathews' work would become immediately relevant at some point, and the H1N1 outbreak of 2009 brought it sharply into focus. If Mathews, a Professorial Fellow at the University of Melbourne's School of Population Health, can solve the puzzle, he may place the world in a far better state to prepare for future pandemics.

'In the 1918–19 pandemic, mortality was greatest among previously healthy young adults, when normally you would expect that elderly people would be the most likely to die,' Mathews says. 'We don't really understand why children and older adults were at lesser risk. One explanation may be that children were protected by innate immunity while older people may have been exposed to a similar virus in the decades before 1890, which gave them partial but long-lasting protection.'

Another unusual feature of the pandemic was that in England and Wales it occurred in three waves, first in the summer and autumn of 1918 and then the following winter. Other countries also had multiple waves, although Mathews says the UK's were 'the most distinct'.

These three phases, along with excellent records as to who got sick and when, have enabled Mathews to model possible explanations for the pandemic's behaviour.

Mathews' interest in epidemiology goes back to his undergraduate medical studies at the University of Melbourne. 'We had lectures from Macfarlane Burnet [one of Australia's greatest scientists], and I read his book *The Natural History of Infectious Disease*, which foreshadowed a lot of the issues people have taken up subsequently.'

Most of Mathews' career has been in epidemiology, and from 1999–2004 he was senior adviser in public health to the Commonwealth government. The period covered the SARS crisis (2003) and the first fears about avian flu in

2004. 'Since leaving I have opened up this line of inquiry with a young group of colleagues to try and squeeze as much data as possible out of the historical records,' Mathews says.

When opening a copy of the UK Ministry of Health's report on the attack rates for the 1918 flu pandemic in the University of Melbourne Library, Mathews says he found a sheet of paper inside with notes in what he recognised to be Burnet's hand-writing. 'It provided me with an extra fascination to pick up this data and apply the latest modelling tools,' he says.

Prior to Mathews' work it was suspected that even with a virus as apparently novel as the 1918 outbreak some people would have had partial immunity based on exposure to ordinary seasonal flu. This explains why the disease was most devastating in isolated areas. In parts of Alaska, where most of the population had not been exposed to seasonal flu outbreaks, more than 80% of the adult population died. Appalling as the death rates were in major cities, they never approached these levels, even in India where overcrowding and poor nutrition drastically exacerbated the crisis.

'When it took many days to get to Alaska by boat, someone who boarded with the disease would not still be infectious at the end,' Mathews says. 'Flu could only reach these locations if many passengers onboard were susceptible and they gave it to each other. On a small boat there would not be enough people susceptible to seasonal outbreaks for the infection to make it.' Consequently, before jet travel, only strains of the virus to which most people were susceptible ever made it to such isolated places.

In other parts of the world, slow internal transport may have blurred the three waves. With these factors removed it is possible to test transmission models based on differing levels of pre-existing immunity across different age groups, taking into account that city-dwellers had more exposure to past outbreaks.

The three distinct phases give Mathews the opportunity to study the theory that short-lived protection from a less serious strain of flu enabled some to avoid the first and occasionally second waves, only to be struck down later as the immunity waned. He has contrasted this with a 1971 H3N2 outbreak in which 96% of residents on the remote South Atlantic island of Tristan da Cunha became infected.

Mathews says his work indicates that 'infection with any sort of seasonal flu provides some protection against pandemic flu, at least in the short term, and the same probably goes for vaccination against seasonal flu'. Recent evidence bears this out. The novel H1N1 pandemic virus of 2009 originated out-of-season in the northern hemisphere and spread into southern Australia during the southern flu season, displacing the seasonal flu virus. Yet in most of Australia, as in other countries, the pandemic virus spread slowly, and only a small minority of people were affected. This suggests that many people were protected, arguably because of prior exposure to seasonal flu, or prior vaccination. However, of those people attacked, a very small proportion became seriously ill, and almost 200 people died from influenza in hospital in Australia in 2009. The attack rate was greater for young adults. Pregnant women, Aboriginal people and those with chronic disease were at greatest risk, presumably because of lesser previous exposure to seasonal flu, and because their immune response to the new virus was less effective. Children were affected, but less likely to become severely ill, and the attack rate in older people was low, presumably because of their greater past exposure to similar forms of seasonal influenza.

In preparing for a pandemic flu, the Australian government contracted for over 21 million doses of pandemic vaccine in 2009. By the time it arrived, the 2009 outbreak had largely petered out, so vaccine uptake has been poor. Nevertheless, the pandemic vaccine was found to give protective immunity after only a single dose, presumably because it was building on pre-existing immunity from the seasonal form of H1N1.

By the time of the 2010 flu season, the Australian population will likely have become more susceptible because of fading immunity and minor changes in the virus, allowing the pandemic virus to make a comeback, possibly with higher attack rates and more severe disease. 'There is thus a real incentive for people to be vaccinated before this happens. Widespread vaccination might save Australia from experiencing higher mortality in the second pandemic wave, as was seen in 1918–19', said Mathews.

'The experience in the 2009 pandemic confirms the idea that the attack rate and severity of influenza is critically dependent on the state of prior immunity in the population, even if the virus is an ostensibly new (pandemic) strain', said Mathews. 'It also raises important questions about the nature and duration of that prior immune protection, presumably induced by previously circulating strains of seasonal influenza. But that is the nature of research – one question leads to an answer, but usually to more unanswered questions – a never-ending story!'

Neuroscientists

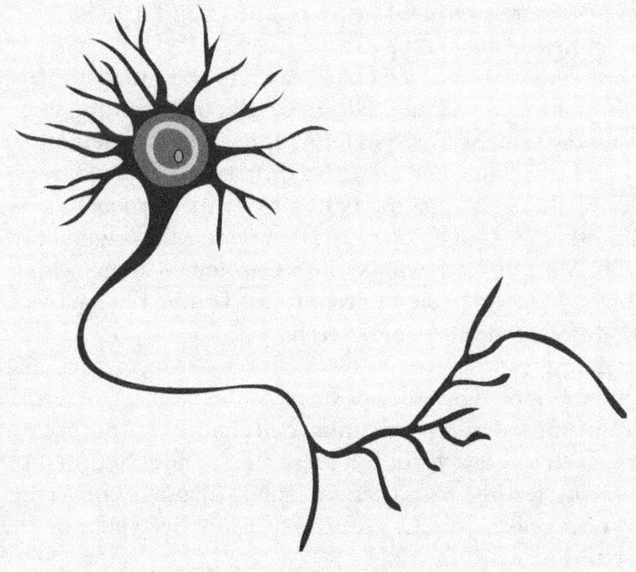

The brain and the Bomb

Most people prefer not to think of radioactivity and brain cells going together, but Dr Kirsty Spalding is combining the two to answer one of the central questions in modern neurology – what parts of the adult brain generate new cells (neurogenesis)? The answer could help regeneration for those who have suffered brain injuries or strokes.

'In the last decade there is a lot of research to show that animals form new neurons [information processing cells] in adulthood,' Spalding says. 'However, it is hard to do the same research in humans because the methods are toxic or dangerous. We needed to come up with a different methodology.'

The idea Spalding is applying is to look at the carbon-14 (^{14}C) levels in DNA. Unlike other parts of living cells, DNA retains the original carbon atoms that formed it. Most of these will be the ^{12}C isotope – atoms with six protons and six neutrons. However, small proportions will be radioactive isotopes of carbon with different numbers of neutrons.

Carbon dating can be used to establish the age of organic material over periods of hundreds or thousands of years, as the radioactive forms of carbon slowly decay. In normal times such unhurried changes are not accurate enough to date a brain cell to a few years, but these are not normal times. In the 1950s and early 1960s, testing of nuclear weapons sparked conversion of large amounts of nitrogen-14 to ^{14}C, creating a distinctive spike in ^{14}C levels in living materials formed at the time.

'For 4000 years ^{14}C levels in the atmosphere have remained (almost) constant; following nuclear bomb testing they rose sharply,' Spalding says. 'After the test ban treaty in 1963 there is an exponential decrease, not because of radioactive decay but because of ^{14}C mixing with large marine and terrestrial carbon reservoirs.'

'In the cerebellum [a region of the brain important in sensory perception and coordination] and a variety of areas of the cortex [the outermost part of the brain] we found no evidence of long-term stable integration of new neurons,' Spalding says. 'Some may be being born, but do not integrate or survive in significant numbers.' This suggests it will be harder to replace damaged neurons in those areas than if the brain already had mechanisms that could be enhanced.

Spalding points out that the results do not contradict previous research. Most animal studies also failed to find neurogenesis in these parts of the brain. 'Some studies claimed to find it, but they are hugely contested, and there are at least as many scientists who argue it is not happening,' says Spalding.

There is preliminary evidence, yet to be published, that suggests neurogenesis does occur in other parts of the adult brain, in line with more widely accepted animal studies.

If Spalding is able to prove adult neurogenesis anywhere in the brain it will crown a dramatic reversal of theory. 'There was a small handful of scientists trying to say there was adult neurogenesis as early as the 1960s,' Spalding says. 'They were largely ignored; until very recently it was dogma that this was not possible. In fact one left science because he was unable to convince others of his results.'

Spalding herself learnt at university that new neurons could not be formed in adults, but the mood started to change gradually as the animal studies became more thorough and stem cell[1] research advanced.

Spalding has expanded this work to look at fat cells. Her conclusion is that the body generates and destroys half its fat cells every eight years. This has the potential to revolutionise research into weight loss, which has previously relied on the assumption that once a fat cell is produced it doesn't die until its owner does. The idea that we destroy a portion of our fat cells every year opens up the possibility of finding a way to speed up the process.

Other scientists have seen the potential of Spalding's work and jumped on board, with colleagues showing the human heart produces new muscle cells, but does so very slowly.

In primary school Spalding read a book called *Know Your Body*, and was fascinated by the section on the brain, copying it out word for word. Upon leaving primary school she announced she wanted to be either a brain surgeon or an actress. 'I lost this focus in high school, but a number of events at the university level, including a friend becoming paralysed, reignited my interest.' Spalding says she was 'frustrated by the lack of treatment possibilities'.

In her Honours and PhD at the University of Western Australia, Spalding looked at how to keep neurons alive in the event of injury. 'We found ways we could delay death, but we couldn't stop it happening,' she says. 'There were a lot of reasons for this.'

Spalding thought she would go to the United States to do her postdoc, saying 'I spent a little of my PhD time at Harvard in Boston and fell in love with neighbouring New York, so was convinced I would head back there for my postdoctoral fellowship.' However, on the way she heard a fascinating talk from Dr Jonas Frisén at the Karolinska Institute in Sweden and asked if she could visit.

1 Stem cells are cells that can renew themselves and turn into different sorts of cells.

Spalding set about trying to study the formation of brain cells in zebra fish. Zebra fish are a model organism, preferred by biologists because they are easy to work with. What Spalding didn't know was that laboratory suppliers specially breed fish that are suitable for study. When stocks ran low Spalding went to the local pet shop and bought more, rather than waiting for replacements. Her experiments didn't work as a result, but her can-do attitude attracted Frisén's attention and he suggested she tackle the neurogenesis research other postdocs had rejected as too much of a long-shot.

Besides shooting her to the forefront of debates about the brain and fat development, the project has unexpectedly led Spalding into forensics. After the Boxing Day tsunami of 2004, rescue teams were faced with the distressing job of trying to identify thousands of victims, many of them with all identifying traces swept away. Spalding showed her ^{14}C techniques could be used to tell the age of a body to an accuracy of 1.6 years (now improved to 6 months), rather than ± 10 years as had been the case before.

Not only did this help establish the identity of many of those who had died, it provoked interest among police investigators confronted by corpses they can't identify. Sweden's Chief Medical Examiner has predicted previously intractable cases will soon be solved as a result.

'I meant to go to Sweden for 10 months, but it's now been almost nine years,' says Spalding. However, with her family in Australia she is attempting to build opportunities for collaborations here. She was interviewed on a trip home made partly to check on local developments.

'It's particularly attractive to come back to Australia now, when I'm sitting on Cottesloe Beach, looking out to Rottnest and knowing it is snowing in Stockholm.'

Nevertheless, she is not keen to give up on the work she is doing. 'The implications are huge. We are living in the era of regenerative medicine. Everyone is curious about the potential of stem cells. There is much interest and hope and still much to learn about the basics of stem cells in the brain and their potential use in therapy.'

The brain collector

Professor Clive Harper collects brains, about 550 to this point. It may sound grisly, but his work in neuropathology and as director of Sydney's Brain Bank has made him a Member of the Order of Australia, and led to a major step forward in public health.

Harper describes his career entry as 'serendipity'. His father was a CSIRO physicist, and pushed him to study physics and chemistry at school. Harper preferred biology and medicine. However, his father doubted he had the intelligence for medicine and enrolled him in dentistry at the University of Sydney.

Armed with impressive first year results, Harper cut the apron strings and transferred, eventually specialising in pathology. Nevertheless, he was yet to find his calling. He met someone who knew a neuropathologist in Scotland looking for an assistant and, 'Keen to see the world, I put up my hand.'

'I loved it,' Harper says. 'I never questioned that it was the right path.' After time in Glasgow and Perth Harper was offered leadership of the neuropathology department at his old university.

Alcohol has been Harper's main field of research, particularly the effects of over-consumption on the brain. He demonstrated that long-term alcoholism leads to a reduction in brain volume and numbers of nerve cells, and that this is focused on certain areas of the brain, rather than spread evenly.

Harper also studied the effects of thiamine (vitamin B_1) deficiency resulting from alcohol consumption. Alcoholics often neglect general nutrition, but this is particularly problematic for thiamine, because 'alcohol actively suppresses thiamine absorption from the gut,' Harper says.

Despite the B vitamins in our national spread, Australians are particularly prone to Wernicke-Korsakoff syndrome, an alcohol-related thiamine deficiency. Harper is not sure why this is, although preference for beer over other alcoholic drinks has been hypothesised. Harper demonstrated the extent of this problem, and in 1991 led the successful campaign to get bread flour fortified with thiamine.

When Harper started his research it was unusual for Australian alcoholics to have problems with other drugs. 'They were a clean sample compared to American alcoholics,' Harper says. US researchers were consequently keen to collaborate with Harper, and he received considerable funding from the American NIH.

'My American collaborators encouraged me to set up a brain bank,' Harper says. Along with one in Brisbane, Sydney's remains one of only two banks in the world specialising in the brains of alcoholics, although Harper admits the pool of people without confounding addictions is shrinking.

Alcoholism is a particularly difficult area for specialisation, because, Harper says, 'You can't just look at a brain and make a diagnosis.' Nurses have to conduct interviews with the deceased's family members and doctors to establish drinking behaviour.

While alcohol was the original motivation for the bank, it has expanded to individuals suffering schizophrenia, bipolar disorder and motor neuron disease. Expansion into the brains of people with Parkinson's and Alzheimer's disease has also begun, and recent funding from the Multiple Sclerosis Foundation means the bank will become the repository for their studies.

It is important for any brain bank to have a significant store of control tissues, and the bank runs the Using Our Brains program through which healthy individuals are encouraged to commit their grey matter to science after death.

Each donation requires far more work for the bank than a simple trip to the morgue, however. 'We have a nurse visit every individual every year to give them memory and learning tests,' Harper says. Consequently, while the bank is keen to get more donors, they would not be able to handle another rush of the sort generated by high profile donors at their 2002 launch.

Many scientists are keen to study samples from Harper's brains, but fortunately the bulk of the research has moved from nerve cell counts to molecular studies. These require much smaller samples, and improved technology has 'reduced the material required ten to a hundred fold.' Consequently it is possible for many researchers to study the same sections of a particular brain.

Eight years after his success in winning thiamine fortification Harper published an article showing that the rate of Wernicke-Korsakoff syndrome had fallen approximately four-fold. Shortly afterwards he was tipped off that some governments were thinking of abandoning the program because 'they didn't think it was working'. Harper was furious that no one had bothered to contact him, but 'fortunately we had published the research' and the program continues to this day, recognised by his 2007 AM.

Harper's stellar career has amply vindicated his choice of medicine. 'When I was offered the position to head the neuropathology department my father wrote me a beautiful letter congratulating me,' Harper says. 'He said he had always wanted to study medicine, but his own parents had not been able to afford it. He was delighted in my success.'

Music to deaf ears

Dr Bryony Coleman is working to take one of Australia's most famous inventions to a new level. Her research could give a new quality of life to the 120 000 people with bionic ears, enabling them to hear voices more clearly and even enjoy realistic-sounding music.

Commonly referred to as cochlear implants, bionic ears use electrical stimulation to activate nerve cells in the ear, enabling people who are profoundly deaf to hear. While cochlear implants have improved in quality and wearability since the first bionic ear in 1978, many users do not find them fully functional and a minority report lower quality of life than before the implantation. In particular, many wearers struggle with conversations in noisy environments or to hear music, particularly a piece they have not heard before.

Coleman's aim is to improve these outcomes by combining the implant with stem cell therapy.[1] Normal hearing depends on sound being detected in the inner ear, where tiny 'hair cells' turn it into electrical signals that are transmitted to the brain via the auditory nerve. Damage to the middle ear can leave the auditory nerves unstimulated, and lack of stimulation causes the nerve to die. Coleman's work aims to use stem cells to rescue damaged auditory nerve cells so there is a stronger channel between the implant and the brain.

Based at the University of Melbourne's Department of Otolaryngology, the promise of Coleman's recently completed PhD won her one of six Victoria Fellowships for young researchers, and the Dean's Prize for excellence in a PhD thesis. Nevertheless, she admits what she is doing is 'in the proof of principle phase and a long way from clinical trials, with many obstacles still to overcome'.

Coleman's first step was to generate a large number of auditory-type mouse cells in a dish and examine their similarity to the functioning auditory nerve in a human. The next step will be to attempt to combine the cell transplants with a cochlear implant to see if they can function normally within the

1 Stem cells have the potential to turn into other cells, and it is hoped they can be used to replace damaged cells in various organs.

body. The most remarkable thing about Coleman's work is that she reached the cutting edge while working on her own. 'There are two places in the world doing this kind of research,' she said when first interviewed. 'The other is at Harvard, where there is a team of 15.'

It takes a fair amount of courage to take on 15 researchers at one of the world's leading universities, even if you are in the old department of the bionic ear's inventor, Professor Graeme Clark. Not surprisingly, Coleman says at first she 'found it intimidating'. At one point Melbourne University gave her a scholarship to present her results in America. 'I thought it wouldn't be seen as good enough, but people were really keen on what I had done and wanted my input on their work,' she said. 'I came back inspired. By the end of the PhD I was a world expert.'

As a result of the Victoria Fellowship's study mission, Coleman now collaborates with the Harvard team, and with Dr Mirella Dottori in Melbourne.

So far Coleman's research has been on animal embryonic stem cells so she has avoided the controversy associated with using cells collected from human embryos, something which has bedevilled many other stem cell researchers. However, successful transplants will almost certainly involve using human cells, and one of the things she used the Victoria Fellowship to do was spend time in California learning about the embryonic stem cell debate. Coleman is now investigating the use of induced pluripotent stem (iPS) cells in her work. IPS cells are seen by many as a way to achieve the benefits of embryonic stem cells without the ethical concerns. However, it is still debated whether they can fully replace embryonic stem cells.

Prior to her current research Coleman was based at the University of Tasmania, where she focused on neurodegenerative diseases (where brain or spinal cord cells decay), such as Alzheimer's. 'During my time at UTas I learnt the basics of neurobiology and this sparked an interest in research into neurodegenerative diseases and exciting new methods by which they may be ameliorated,' Coleman says.

Coleman was drawn to cochlear implant research as it was an important opportunity to work in medical bionics – improving the interface between a neural prosthesis and the brain. 'Here we have a device which works, but can be improved,' Coleman says. 'And stem cell science is an exciting emerging therapy for neurodegenerative disorders, with lots of potential for new discoveries.' With Australia's track record in cochlear implant research, Coleman enjoys being able to communicate her results readily to scientific audiences and the general public alike.

'The absolute gold standard would be to regenerate a fully functional auditory system, but initially you'll still need an implant because you lose the hair cells first, then the neurons. Each hair cell responds to a particular frequency. The implant replaces the hair cells, transmuting sounds to 22 electrodes stimulating the relevant neurons. The problem is that the nerves keep dying, compromising the effectiveness of the implant.'

The bionic ear was an unusually controversial medical treatment, generating powerful antagonism from members of the deaf community who saw it as treating deafness as a disease rather than an identity. However, Coleman says she has received none of this criticism, perhaps because her work is designed to help those who already have an implant.

Coleman remembers as a child growing up in rural Tasmania and seeing her grandfather chop the head off a snake. Four frogs were inside, and one of these hopped away. 'This sparked a lot of questions,' she says, and may have been the trigger for a lifelong fascination with biology.

Having gone through school with 'no idea what I wanted to do, but it had to be something related to biology', Coleman studied biochemistry at university and was offered a place in the Honours program at the University of Tasmania.

After this Melbourne offered 'a fantastic team full of people from multiple disciplines; engineers, biologists and electrophysiologists' as well as daily exposure to patients who may one day benefit from her work. Coleman describes winning the Victoria Fellowship as 'a career accelerator. It has brought lots of attention to my work. A lot of getting funding is people knowing you. Publications alone aren't enough any more, and this will help me stand out.'

Since being interviewed for the first version of this profile, Coleman has received a four-year fellowship from the National Health and Medical Research Council of Australia to investigate the potential of stem cell therapy for cochlear implant recipients. She is also supported by a Wagstaff Fellowship in Otolaryngology from the Royal Victorian Eye and Ear Hospital. Coleman was recently married, and some updated information is available about her under the name Bryony Nayagam.

Her professional profile can be viewed at www.medoto.unimelb.edu.au/research/hearing_preservation/stem_cells_and_regeneration. She is currently establishing her laboratory in Melbourne and keen to announce she is looking for talented students to help continue her research.

Music and the mind

On his next sabbatical, Professor Alan Harvey plans to write a book about the evolutionary importance of music, and why and how it can have such a profound impact on our moods and call forth such deep responses. It is a book he is especially qualified to write because few other neuroscientists have also achieved success as musicians.

Harvey has played support for giants of the folk music scene (including The Fureys, his cousin Eric Bogle, and Maddy Prior of Steeleye Span). Harvey says, 'I hope the book will be a combination of neuroscience, evolution, human biology and art, finally bringing all the strands of my professional and other lives together.'

Harvey's research is primarily on degeneration and neurotrauma (damage to the nervous system), spinal cord injury, and cell replacement and pathway repair in the visual system. Gene therapy has been a particular focus of his recent research work. He is a member of the International Neural Transplant Council, the Neurological Council of Western Australia, and for many years he was a member of the WA Reproductive Technology Council. He is Chair of the International Programming Committee for the bi-annual Asia–Pacific Symposia on Neural Regeneration.

Working at the cutting edge of a field where people are desperate for progress, he says 'Over the years I have learnt not to make predictions about when problems will be fixed.' He does not, in the short-term at least, expect to see a dramatic cure enabling those in wheelchairs suddenly to walk. Instead, he expects 'gradual but sustained progress through a convergence of new research: using physiotherapy as well as transplants and advanced molecular biology. There is evidence that physio can reawaken things that have shut down. There is a move afoot to bring laboratories together to combine all these approaches: physiotherapy, cellular transplants, gene therapy, immunotherapy etc.'

Music also runs deep for Harvey. He studied piano for 6 months at the age of 8 'and then my teacher suddenly died and I wouldn't have lessons from anyone else. So I basically taught myself from then on.'

Teaching himself to play guitar and mandolin while doing Bachelors and Masters Degrees at Cambridge led to a folk–rock band. Harvey was invited to play with another band, the drummer of which went on to considerable success.

Had he accepted his career might have been different. However, science won out as he had already accepted an offer of a PhD in visual neurophysiology (the study of the functioning of the nerves that enable us to see) at the John Curtin School of Medical Research in Canberra. An invitation to audition for a band while doing his post-doc in Seattle also had to be turned down, although he did find time to sing in the Seattle Symphony Choir.

While Harvey's scientific career has taken him to a chair at the University of Western Australia (UWA), he's performed solo, played in various folk bands, and now plays keyboards and guitar in a rock'n'roll band called Chain Reaction, as well as assisting occasional projects with friends. Asked if he ever considers leaving science for music full time, he replies 'I sometimes dream of retirement, but I'm not sure if I'm good enough to be a professional musician. It's easier to be a scientist and part-time musician than the other way around, particularly because science requires so much institutional research and is so complex and fast-moving.'

Nevertheless music has been 'an important outlet, a safety valve' for Harvey. 'Even if I'm down before a rehearsal I'll be smiling afterwards.' He points out that 'medical research and science are full of people with creative skill as a second string. Most scientists are interesting people.'

Teaching is also important to Harvey as well, but unlike Tom Lehrer, who combined folk singing with lecturing in mathematics at Harvard, he does not encounter students who expect him to sing in class. 'Most students don't know,' he believes. 'Even most staff in the faculty. Occasionally we have a good gig, like when we supported The Fureys and there were 2500 people there. Later, some medical students said, 'We were certain we saw you at the concert last night – was that really you?' They were surprised the boring old academic who teaches them basic cortical anatomy shakes his booty on stage.'

Nevertheless, Harvey believes his musical career has improved his teaching. 'There's no doubt that performance and experience overcoming nerves helps me lecture.' But the skills required on stage are relevant too. 'As a musician, especially in a band, you learn to listen to others while performing. You need to be responsive to everyone else. When I give a seminar I'm more aware of the audience and their level.' Harvey has a Science Teaching Excellence Award and Research Supervision Award to prove how well his stage experience has paid off.

Harvey can't explain what led him to science. 'At 9, I wanted to be a great composer, but "great" was the main thing!' His mother told him he expressed an ambition to be a medical researcher at 12, but he has no recollection of the fact. 'I guess it was just inevitable. [At school] I didn't like languages, but I liked biology. When I got to Cambridge the Department of Physiology was so good you couldn't fail to be stimulated by the fine minds there.'

Physicists

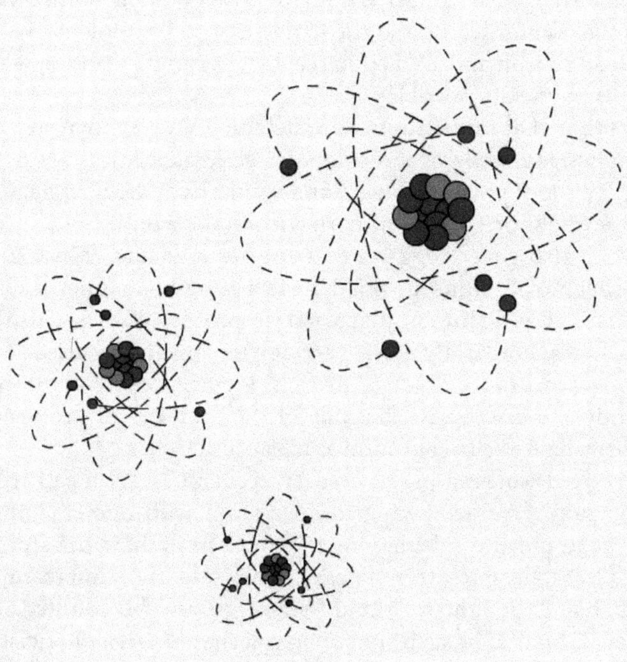

The stunt physicist

If Hollywood ever wants to make a TV series making physicists look cool, the way *Numbers* tried to do for mathematicians, they could do worse than consider Suzanne Hogg. The danger is it might give too many of their tricks away. Until recently a senior lecturer at the University of Technology, Sydney, Hogg also consulted for film and TV action scenes, advising TV and movie directors about what is physically possible.

'The truth is that most stunts we watch on TV or a movie are real. Computer animation is expensive and computer-generated stunts are not exciting,' Hogg said. 'Without physics James Bond would have died long ago and there would be a severe shortage of Aston Martins in the world.'

The work also gave Hogg status with her students. 'Most students are impressed that movie stunts are not done in a gung-ho manner and that physicists are consulted as an important part of the process. They are also impressed that studios are scientifically literate enough to consult physicists. In fact, studios cannot get insurance unless a physicist is consulted for stunt work,' Hogg said. Recently she discovered that one of the founders of professional stunt work in Australia had a background as a physics teacher.

Hogg started working in the area by accident. 'I got a call from a stunt person who said: "We are trying to do a stunt with one car jumping over another".' Hogg did the calculations to demonstrate how the stunt could be performed. 'My colleagues were appalled. They said [the stunt team] might kill themselves, but I thought we should stand up and be counted. I was also pleased they knew that in such situations you should call a physicist.'

In order to increase the safety of the operation, the stunt agency provided Hogg with real-life data from similar stunts, which she fed into her model to confirm her calculations were correct.

While the stunts are photographed at angles that make them appear more dramatic than they really are, Hogg claims most are highly realistic. She notes Australian film budgets usually don't run to repeat takes, and in one case she was told 'one brand new Audi had to jump over another and not have a scratch on it'.

In another case Hogg had to calculate the take-off speed and angle required for a motorcycle stunt rider to travel a path matching the curve of the

Sydney Harbour Bridge, which formed the background to the scene. The stunt rider took her advice, but demonstrated her own expertise by not using a speedometer. 'She had to be looking where she was going to land – she established the right speed by ear, listening to the sound of the motor,' Hogg says.

She regrets the intermittent nature of the work, and that now she is retired from UTS she can't afford the insurance required to continue to consult for stunts. 'I love working with stunt people because they are so professional and love their work.' The work is too unreliable for her students to do work experience with a stunt agency, despite enthusiasm from both sides.

Hogg's own entry into physics required some luck. She only gained permission for a late transfer from some humanities subjects on the basis that her brothers, both physicists, would be able to assist her. After a Bachelor of Applied Science (Honours) from the University of Western Australia she did a Masters at UTS. During her career at UTS Hogg investigated neutron diffraction and crystallography while teaching mechanics, electromagnetics and quantum physics. She is an advocate of using audiovisual demonstrations in physics lectures, rather than relying on the old style chalk and talk.

In the latter part of her career Hogg focused on acoustic research, an area that fits well with her 'night job' as a concert pianist. 'When I finished university I had to choose between being a physicist and a musician, and I decided I preferred to play music when it suited me rather than when someone else wanted.' She has maintained the music under her maiden name of Maslen, and has four musical children. 'I've got a string quartet, a vocal quartet and a brass quartet out of my family.'

It hasn't always been easy. Attempts to combine the two disciplines at university forced her 'to get up very early each morning to study [physics] and practise [music] late into the night'. Despite the wealth of musical physicists in this book, Hogg found that in one job 'the culture there was very much against admitting that you were going to do anything else over the weekend other than more physics,' so there have been advantages to keeping her two careers under separate names. However recognition came in the form of an Australia Day Medal of the Order of Australia.

Hogg uses examples from stunt work to boost enthusiasm for physics among many different types of people. 'I get the same response from Rotarians as university physics students, but the questions they ask are different. The students ask things like, "Did you allow for the drag coefficient at that point?", while the Rotarians want to know why [the stunt rider] didn't use a speedometer.'

The man with X-ray vision

If scientific awards attracted the same publicity as sporting trophies, Professor Keith Nugent would be one of the most famous names in Australia. Few scientists in the country can dream of winning the honours he has received, but he's not bothered that fame hasn't followed.

Nugent's most celebrated discovery related to his observations of the phase of light.

Because it is a wave, light has phases, which means interacting light waves can be thought of like two ocean waves meeting. Light in phase combines like two waves coming together to produce a major crest, while when out of phase one wave's trough cancels out another's peak.

Interference patterns from out-of-phase light beams have been one of the key experimental tools of the twentieth century, but their use has relied on sharp, coherent[1] sources of light to start with. It's relatively easy to use the interference fringes produced by a laser to work out the shape of the object it is shining on, but it is much harder to do so with a large source emitting a range of wavelengths.

Nugent showed that it is possible to use the shifting bands of light on the bottom of a swimming pool to work out what is happening on the surface using a very large, very incoherent source – the Sun.

Nugent says the work has had applications in microscopy where you can 'make a measurement shining light through an object, defocus it a little and observe the changes. You get a robust phase[2] measurement of its shape.'

In 1999 Nugent established the company Iatia to commercialise these techniques, which are now used around the world, for example in non-destructive cell counting. 'Cells are transparent. If you stain them [so they can be seen more easily] they die,' Nugent says. 'Using phase contrast techniques you can count them quickly and reliably without staining.'

The work has led to the establishment of an Australian Research Council Centre of Excellence at the University of Melbourne, where Nugent is based.

---

1 Coherent light sources produce light of a single wavelength or a small band of wavelengths that can be directed into a very precise beam.

2 The phase of a wave is the fraction of its cycle it is away from some reference. Phase measurements combine light in different phases to produce a clear picture.

The primary goal, Nugent says, is to develop the techniques to the point where the exceptionally powerful X-rays from a synchrotron can be shone on a single biological molecule that cannot be crystallised, and used to reconstruct its molecular structure. In the absence of such a technique it is extraordinarily difficult to know the structure of such molecules, and without knowing the structure it is very difficult to design molecules that lock on to them.

Nugent also developed the lobster-eye X-ray telescope with Dr Steve Wilkins of CSIRO. 'The X-ray sky is dynamic,' Nugent points out. 'You want to be able to monitor large areas of the sky for developments, so when something happens you can quickly point a high resolution telescope at it.'

X-rays are the part of the electromagnetic spectrum with wavelengths between 0.01 and 10 nm (visible light is 390–750 nm). Other parts of the electromagnetic spectrum represent less of a problem, and not only because they are usually less prone to sudden developments. In the optical wavelengths (visible light) it is possible to construct fish-eye lenses to scan large areas of the sky at once, but X-rays cannot be refracted with a lens so a mirror is required. The atmosphere blocks X-rays, so an X-ray telescope needs to be in orbit, making the use of multiple scopes impractical. Nugent's solution is modelled on the eyes of lobsters and prawns. It uses multiple corner cubes[3] with one face removed so that reflection occurs in two directions rather than three. A curved array of these cubes reflects a parallel beam to a focus like an ordinary mirror. A large enough array can pick up developments across a huge area of the sky at any one time with sensitivity approximately 1000 times previous efforts.

The work has excited X-ray astronomers, and was chosen to go on the International Space Station in 2007. Regrettably Nugent says the station is 'way behind schedule because of problems with the space shuttle. They keep blowing up, bits falling off …'

Nugent always expected to be a scientist, but an enthusiasm for bird watching originally made him lean towards the biological sciences. However, in Year 11 his school made a trip to central Australia. Two weeks sleeping under the stars excited an interest in astronomy, an enthusiasm bolstered by the discovery that he was good at maths.

An undergraduate degree in physics at the Australian National University (ANU) led to Honours in atmospheric science at Adelaide University. However, Nugent found atmospheric work at the time 'very observational and unstructured'. Keen to make a contribution to society, he headed back to ANU to do a PhD in laser fusion, an idea to produce energy by forcing hydrogen atoms together using lasers.

Nugent realised that laser fusion was never likely to become an economical source of clean energy, and that research in the area was mainly useful for

3 Corner cubes are cubes with one face removed. The other faces intersect so that waves coming in at any angle are reflected back to the source. Examples include roadside reflectors.

nuclear weapons. This was 'the exact opposite of the reason I got interested in it'. However, lasers sometimes produce X-rays when they strike objects, and examining these led him in more constructive directions.

In the course of his work Nugent has won the Victoria Prize and an accompanying Anne and Eric Smorgon Memorial Prize. He is the only Australian to win two R&D 100 awards (for the 100 most significant technical innovations of the year) and holds 10 patents. He is a former head of the University of Melbourne's Physics Department and deputy chair of the National Scientific Advisory Committee to the Australian Synchrotron Project. Among other awards he has won the Pawsey medal for a physicist under 40, the Boas Medal from the Australian Institute of Physics, and has twice been awarded Federation Fellowships, a program designed to keep Australia's leading researchers at home by employing them on pay and conditions that could previously only be obtained overseas.

Superconducting physicist

As President of the Federation of Science and Technology Societies (FASTS), Dr Cathy Foley leads an organisation that promotes science to the government and community and offers services to 60 000 working scientists. She is also the immediate past president of the Australian Institute of Physics (AIP) and was the first woman to hold the position in the Institute's 48-year history.

While the two-year presidency is not a paid position, her employer, CSIRO, does see it as part of her job, enabling her to attend conferences and work the FASTS and AIP roles into her employment as Research Program Leader with CSIRO's Materials Science and Engineering Division. At Materials Science she is working to develop superconducting sensors that can detect undersea oil without interference from ocean waves, detect magnetic ore deposits from aeroplanes, and detect the reversal of spin direction of a single atom.

Leading the program takes up the largest share of her time, and Foley estimates that only about 30% of her work time these days is spent on her own research. She keeps up with other progress in her field from home after hours.

While she admits to a certain frustration at this, she also says: 'Research requires organisation and management. A lot of scientists are lousy at these. I think I have reasonable people skills and perhaps there's an obligation to use them. On a personality test I'm probably the exact opposite of what you'd expect from a physicist.'

It's not all sacrifice either. 'I enjoy working with people to make them the best they can be. I see myself as having a multiplier effect by creating a good research environment. Sometimes I'd like to be in there doing it myself, but I enjoy seeing people become fantastic scientists in the environment I've created.'

There are 130 scientists in her program, so Foley has plenty of opportunities for that, although she adds that she has a considerable support structure. Foley 'didn't come from a family of scientists' and says she's 'not sure what drew her to science'. However, she was fascinated by nature studies at primary school and excited at the prospect of studying science when going to high school. She entered science competitions at school but didn't win any, other than one '$7 prize'.

Science appealed as a career, but Foley says: 'I didn't know any scientists.' Research didn't look like an option, and instead she thought she'd become a high school science teacher. At Macquarie University, however, her second-year physics lecturer held barbecues at his house and 'made me feel welcome'.

Foley realised that if she could get a scholarship she could do a PhD. She 'ditched the education department', winning first class Honours and the scholarship. Her thesis was on nitride semiconductors, a field she says has 'since taken on a life of its own' through a very successful spin-off company. In the course of her work she grew the world's purest indium nitride crystals, a record only recently broken after 25 years.

From there Foley went to work with the then-new field of high-temperature superconductors. As objects cool down their resistance to electrical flows slowly falls. However, below a certain temperature some substances suddenly become superconductors, where their resistance to a current becomes exactly zero. This isn't just useful as a way to cut the loss of energy in transmission, it also creates bizarre interactions with magnetic fields.

Professor Brian Josephson won the 1973 Nobel Prize for predicting the invention of Josephson junctions, where a superconducted current is briefly interrupted to create an extremely sensitive device for the detection of magnetic fields. At the time of his prediction the only known superconductors required temperatures too cold to make this very practical.

The discovery of 'high temperature'[1] superconductors opened the door to widespread applications of this effect, and Foley has put the technologies together for multiple uses. For example, the exquisitely sensitive detection of magnetic fields can be used to find oil and mineral deposits. To locate such deposits under the ocean you need to eliminate the electrical noise created by surface waves, so Foley is finding ways to operate superconductors at a depth of 1 km below sea level, something she describes as 'a challenge'.

In the course of her work Foley has won the 2005 Eureka Prize for the promotion of science, partly for her six-year run on ABC radio discussing scientific developments in a weekly evening slot. Although this required about 16 hours of unpaid research per week, she was disappointed when a change in presenter saw her ditched for a dream interpreter. Last November, she won the National Telstra Women's Business Award – Innovation.

Foley believes this background has been useful in her roles as president of FASTS and AIP, providing a better understanding of the media. She also

1 'High temperature' in this context means anything that superconducts at more than 30°K, or –243°C. The record for high temperature superconducting is still much colder than an Antarctic winter.

believes she brings a pragmatism to the role that not all of her predecessors have had, including a capacity to 'squeeze things into small blocks of time'.

She is disappointed that the AIP's campaign to address the issue of how few school physics teachers have physics qualifications 'hasn't gained traction'. On the other hand, she has hopes her two-year position on the Australian Prime Minister's Science Council will yield great results.

'Physics gives you a really good critical thinking background,' Foley says. 'It's amazing how many heads of large organisations studied physics. They didn't go into research but got promoted really quickly on another path.' She is hoping the Institute can make employers, and potential students, aware of this fact.

A quantum of music

It's a long way from quantum computing to didgeridoo playing. Professor Lloyd Hollenberg's work covers both, and much in between. What's more, just days before our interview Hollenberg, of the University of Melbourne, discovered the apparently unconnected fields have something in common.

Didgeridoos are among the most ancient technologies still in use today, while quantum computing is a field so new it is more a promise than a reality. Although breakthroughs have been made in using the extraordinary behaviour of very small objects to make extremely powerful computers, a machine that actually works is still years away.

Hollenberg conducts theoretical modelling for the Centre for Excellence in Quantum Computing. Two of his students demonstrated that it is possible to overcome the problems that occur when one tries to get a one-dimensional quantum computer to factor a large number.

'Short's algorithm says that quantum computers could factor very, very large numbers much more quickly than ordinary computers and the factoring problem is important because it lies at the heart of data security,' Hollenberg says. 'However, there were concerns that with a computer made of a line of qubits,[1] where each interacted with its neighbour, scaling [up from a smaller system] might make it difficult when it needs to interact with a qubit a long way away. My students showed it can be done in a very clever way.'

At the same time, Hollenberg has made himself an expert on the acoustics of the didgeridoo (or yidaki, as it is called by the Yolngu people of north-east Arnhem Land). 'I started playing first. At the time I knew nothing about acoustics. I played the piano and I wanted an instrument that was portable, and also wanted to learn more about Indigenous culture.'

---

1 Qubits, or quantum bits, are single pieces of quantum communication. They are the quantum computing equivalent of an ordinary computer's bit. Ordinary bits can only be either 0 or 1, but qubits use the extraordinary properties of quantum mechanics to be 0, 1, or both at the same time. If you don't understand this don't worry. Many scientists suspect that no one understands quantum mechanics, even those who've been working in the field for decades.

However, it was natural that a physicist would start to wonder about the workings of his new instrument. Hollenberg searched the available material and was amazed to find only two published papers on how the sound of the yidaki is produced. He taught himself musical acoustics and got in touch with Neville Fletcher, the author of the first paper, and Joe Wolfe (p. 155).

Since then the understanding of yidaki acoustics has expanded greatly, due to the collaboration Hollenberg created. 'The pipe is important to the characteristics of the sound, but players will tell you that the music really comes from the mouth and vocal tract. It's very difficult to understand how sound is formed in very differently shaped mouth cavities through extraordinarily vibrating lips down a pipe.'

To increase the understanding of what the mouth of a yidaki player is actually doing, Hollenberg organised to have magnetic resonance imaging (MRI) scans done of him playing the instrument. However, this was no easy task. He had to lie flat in the MRI, and 'there is not a lot of room in there'. Instead of a traditional wooden instrument he used a PVC pipe with a bend in it so the body of the instrument could lie next to him.

To get the range of sections required for a 3D image, Hollenberg had to hold a note for 20 seconds while also controlling the MRI from within the machine. Even then the resolution of the MRI needed to be reduced. Nevertheless, he managed to get 'great images' of two notes that 'kick-started our modelling, and helped us create a controllable automatic didge player and understand the measurements we were making'. The results of this Australian collaboration were published in *Nature* in 2005.

Hollenberg puts this knowledge to use in talks on the physics of yidaki playing. These cover a lot of ground, with demonstrations on the physics of sound from the well-known shattering glass to the effect of putting sound waves down a tube full of burning gas. He's even managed to broaden the appeal by working in a connection to dinosaurs after palaeontologists found a skull with a hollow crest connected to the nasal passages. Other scientists worked out the resonant frequencies[2] of the skull cavity and made a sound file of the song of this particular dinosaur, and Hollenberg plays this to audiences.

The talks, Hollenberg thinks, also help build respect for Indigenous culture. He has spent time in Arnhem Land learning from elders about their perspectives on the workings of the yidaki, and gaining the appropriate permission to conduct the research. 'I think it is good for students to hear me as a scientist trained in the Western intellectual tradition talking with respect about the Indigenous view of this instrument. I'm not qualified to talk in depth about Indigenous culture, but I can discuss at a simple level how different clans prefer different shapes of instrument and so on.'

---

2 That is, the notes produced when air is blown through the skull's nasal passages, giving an indication of what noises the dinosaur might have made when alive.

Hollenberg says he was 'always interested in taking things like radios apart and most people thought I would be an engineer'. However, it was not until he reached Year 12 that he began to really like physics.

'I saw a documentary on special relativity and it blew me away,' he says. Science had to compete with gymnastics for a while, and during the final year of school his studies suffered from training six nights a week. On the other hand, the mechanics of gymnastics gave him an appreciation of classical physics. He turned down an offer to study at the Australian Institute of Sport in order to take up a place at Melbourne University, where his Bachelor of Science turned into a PhD and then full-time employment.

And the connection between his two areas of research? To couple solid-state quantum computer qubits to light, scientists have recently used long thin microwave resonators that behave in a manner similar to didgeridoos, albeit only millimetres long.

Physics made fun

Not long before being interviewed for this chapter, Professor Joe Wolfe was awarded two prestigious prizes in a fortnight. One was for research, the other for teaching. Wolfe is Professor of Physics at the University of New South Wales, specialising in the study of sound. However, his teaching ranges over many areas of physics, finding ways to explain concepts that frequently baffle students and can discourage them from continuing their studies.

It was this enthusiastic approach to education that won Wolfe the Physical Sciences and Engineering division of the 2004 Australian Awards for University Teaching. Wolfe created a course called 'Physics Thinking' that, rather than simply teaching physics concepts, aims to get undergraduate students to think like a physicist.

Wolfe's regular physics classes were good enough that students nominated him for university awards in teaching, which led to the national prize. He says he enjoys 'bringing the physics into his classroom' with numerous demonstrations. 'It's great getting a whole lecture theatre talking about a demonstration.'

Wolfe has also created an extensive set of web pages explaining the physics behind a lot of everyday experiences (www.phys.unsw.edu.au/~jw/teaching.html). He says 'a lot of it is written for my students' but he also moderates a bulletin board on the NSW high school physics course, as well as providing web pages for those studying physics at school.

Some sites are designed to appeal to an audience with little physics background. Many musicians, for example, are interested in the acoustics of the instrument they play, and Wolfe has a large site to deal with many of their questions while assuming very little prior physics knowledge.

'Musicians are an interesting audience because they are prepared to put the effort in,' Wolfe says. 'They are not like the average web user. I get messages from people who have worked their way right through the whole site.'

Wolfe's website is also of interest to sports fans. Who could resist a link titled 'the quantum mechanics of cricket' featuring a photo from a 1999 match against Zimbabwe when then-Australian captain Steve Waugh had all nine fielders in slips?

Wolfe says: 'As soon as I saw the photograph I thought "single slit diffraction"'. One of the extraordinary features of the quantum world is that when a photon (or single unit of light) passes through a very narrow slit it doesn't pass straight through. Instead its path will be bent in such a way that it can arrive at certain locations on a screen behind, but not others. The pattern formed in such experiments forced possibly the twentieth century's most important overhaul of our understanding of how the world works.

If it was possible for a cricket ball to diffract between bat and pad, Waugh's field would be the perfect one to maximise the chances of catching the batsman. Wolfe then set a homework problem based on the idea, and included the answer, with jokes, on the site. Unfortunately, as Wolfe proves, in order to get quantum diffraction effects on an object the size of a cricket ball one would need to bowl at a speed of 3×10^{-20} m/s. At such a slow speed, the Sun would have burnt out long before the ball travelled the length of the pitch.

Shortly after winning the teaching award Wolfe collected La Médaille Étrangère 2004 from the French Acoustical Society. This is awarded each year to honour a non-French scientist who has contributed to acoustics and has links with French acousticians. The award is for general achievement rather than a specific piece of work.

Wolfe says that his main contribution is that the lab he co-leads with Dr John Smith 'is really good at measuring acoustic response in noisy environments'. This has produced world-leading research on the how musicians use their vocal tracts in singing and playing. Two spectacular examples are the projects on the didgeridoo (p. 152) and on sopranos, including why they are so hard to understand when they are singing the high notes.

Wolfe says he was 'one of those children who is always asking "why?"' Why, he found, 'often led to physics. Why is the grass green? Because of chlorophyll. Why is chlorophyll green? Because it absorbs certain frequencies of light, which led to quantum mechanics.'

Wolfe says he got good answers as a child. 'My father was a science teacher, and I had several teachers who pointed me to references including the Encyclopaedia Britannica article on space–time, written by Einstein.'

Becoming a physicist was certainly an option as a child, but it had competition, particularly from music. Wolfe is a wind player and composer. One of his strangest but best-known pieces is the Stairway Suite, commissioned by the UNSW Orchestra.

Inspired by Andrew Denton's TV show *The Money or the Gun* (ABC, 1989–90), which featured a different version of 'Stairway to Heaven' in each episode, the orchestra sought an orchestral version. In the end Wolfe provided a suite of six variations on the song, each in the style of a different orchestral composer.

Despite this dual interest, Wolfe did not start out specialising in music acoustics, saying, 'It didn't occur to me that it was possible.' After a Bachelor of Science from the University of Queensland and a Bachelor of Arts at the University of New South Wales Wolfe did a PhD at the Australian National

University. His early research was on the self-assembly of biomolecules› molecules produced by plants or animals, including humans.

'The main thing that holds molecules together in the structure of the cell is the surface tension of water,' Wolfe says. 'I looked at what happens when you take away the water. Examples are seeds and plants in drought conditions, or frost damage and cryopreservation[1] of cells and tissues.' And yes, there was a medal for that work too.

1 Preserving living things by cooling them to very low temperatures. Blood components, sperm and embryos are stored this way. An extreme example is the idea of freezing people with terminal diseases so they could be thawed out in the future when medicine has advanced enough to save them. Most scientists regard this as impossible, but more limited versions, such as freezing organs before transplantation, are being studied.

A quantum leap for children's fiction

Quantum mechanics is so strange that adults are often deeply disturbed by attempts to understand it. So the idea of introducing it to children might seem utterly foolish. But Dr Yvette Hancock disagrees. 'Children have such vast imaginations,' she says, 'and quantum mechanics is so bizarre they may understand it better than adults.'

In 2003 Hancock completed a PhD in physics at Monash University. A few months later the idea occurred to her to translate it into a children's book. Within two hours she had created the character of Ellie the electron who lives in a quantum circus where the tents are atoms, all governed by a Mr Pauli who makes rules about how Ellie and her electron friends can behave (a reference to the Pauli exclusion principle which governs electron behaviour).

The book introduces established, if difficult, scientific theories such as the Heisenberg Uncertainty Principle. But in keeping with Hancock's aim to 'bring cutting-edge research to children,' it explains her own work on switching in small groups of atoms. To make sure the science was right, Hancock had the book 'peer reviewed, including by researchers around the world'.

The idea has clearly struck a chord, with media interest including *The Age*, *Deutsche Welle* and Hancock's local paper. Channel 7 was so interested they filmed Hancock reading her book to a prep class at a nearby primary school. 'By the end the children were saying "electron". I felt quite emotional. I've never seen this kind of response.'

Emails from around the world have streamed in. A few months after the media buzz Hancock got 'five or six a week from around the world', with many asking Hancock where they can buy a copy. It might be expected that publishers would be beating a path to Hancock's door, but as yet the book remains unpublished, perhaps because it crosses boundaries. 'It's not an educational text, it's children's literature. You get drawn in.'

Hancock says that children's literature is 'quite a strange industry', and publishers are so deluged with texts that the best way to be published is to have someone within the industry take up your cause. Hancock has the May Gibbs Children's Literature Trust advocating on her behalf, and is hopeful a publisher will be found soon. Meanwhile she is half-way through a sequel and has plans for a series as Ellie takes us to more and more complex areas of subatomic science.

The second book, which Hancock is halfway through and is proving more of a challenge to write, introduces the character of Psi, the wave function, who mysteriously knows a lot about Ellie and her friends. (A confusing aspect of quantum mechanics is the way objects can simultaneously be particles and waves with behaviour predicted by mathematical expressions called wave functions). As the books go on, more 'mathematical expressions will become characters'.

Hancock does not remember a particularly inspiring book as a child, but she did learn to read at two-and-a-half, which might contribute to her confidence that very young children can deal with the concepts of advanced physics. 'I was an avid reader, always reading several years above others in my class,' she says.

Hancock 'didn't wake up at five and decide to become a theoretical physicist'. Indeed, physics was not a subject she found particularly easy at school.

She also took studies in violin, clarinet and dance seriously, considering careers with each. However, physics was challenging and, she adds, 'My questions took me there'. Indeed, after enrolling in a Science/Engineering course at Monash, Hancock was told by a lecturer that she 'asked too many questions to be a good engineer – perhaps you'd like to consider a career in science'.

If this was not enough to put her off engineering, Hancock describes a feeling of considerable disappointment when the engineering students entered a challenge to build a 'beer can delivery device'. 'Ours didn't work on the day. It worked the night before. The one that won was basically just someone putting it in a post pack and throwing it.'

Science, it seemed, was a better option for the inquiring mind. However, even further twists were in store. Hancock's Honours studies were in experimental physics, but again her questions led her in the theoretical direction.

After the PhD on electron switching, Hancock spent time at Helsinki University of Technology researching nanoscale devices. Her work is now being used by Nokia to design the next generation of mobile phones, but Hancock is pursuing similar topics as a lecturer in the Department of Physics at the University of York in the UK. 'People sometimes ask me if I am going to change careers to become a writer, but I couldn't write these in isolation from my research,' she explains.

To Hancock, the link between physicists and children is natural. 'We do what children do, like ask: "Why is the sky blue?"'

Tennis anyone?

Associate Professor Rod Cross has made a specialty of the diverse fields of sports science, plasma physics and murder investigations.

'Physics and maths were always my favourite things at school and I thought I'd become an engineer,' Cross says. However, his parents couldn't afford to pay university fees and the cost of moving from Forbes to Sydney, so he accepted a teaching scholarship and expected to become a science teacher.

'After four years at university I loved it so much I wanted to do a PhD, and convinced the Education Department that teaching at university was equivalent to school teaching.' A Doctorate in plasma physics led to 30 years studying shockwaves first and then Alfven waves (a sort of low frequency oscillation in plasma[1]).

Cross says his most exciting result was the discovery that it was possible to launch an Alfven wave and have it stick to electromagnetic field lines and 'propagate almost like a laser'. These waves could do several loops of the tokomak[2] he built at Sydney University before they dispersed.

Eventually, however, Australian Research Council funding ceased. Cross's PhD students finished on a combination of small grants, private funding and 'petty cash'. 'I tried getting into more applied aspects of plasma physics, such as plasma deposition, but they wouldn't fund me because I had nothing published in that area,' he says.

Still with a University of Sydney position, but approaching the end of his career, Cross decided he needed to get into something cheap to study. He tried the physics of tennis and 'found it fascinating'. Out of this has come three books and many papers revealing why tennis rackets have sweet spots, the best place to aim the ball, and why it's better to serve from the top of the racket than the centre. Cross also tests the speed and bounce of newly resurfaced courts each year prior to the Australian Open.

Cross has also investigated the surprisingly understudied physics of baseball, and produced a paper that was rated among the 75 best physics papers in

1 Plasma is a state of matter where a gas has been heated to the point that electrons have come free from many of its particles, making it electrically conductive.

2 A machine that uses magnetic fields to force plasma into a doughnut shape.

America for 1998. Only around 10% of athletes whom Cross meets seem interested in his work, but he's had much more success with coaches who are keen to know if their explanations of what works are supported scientifically.

More famously Cross was the key expert witness for the prosecution in the investigation of the death of Caroline Byrne, whose body was found at the base of The Gap in Sydney in 1995. This was one of 'six to eight times' Cross has been consulted in relation to falling deaths, but the Byrne case made headlines for years. In another case Cross demonstrated that an accidental fall was far more likely than murder.

Cross believes that falls are seriously under-researched. 'In the USA, falls cost $37 billion per annum compared with $49 billion for motor vehicle accidents,' he says. 'The cost of stair falls in the USA is the same as the cost of building new stairs. Huge sums are spent on researching motor vehicle accidents but almost nothing on falling accidents since people (e.g. lawyers and architects) tend to blame the person who fell rather than focus on potential design or maintenance issues.'

When police phoned Cross about Byrne's death he immediately dismissed the possibility that she was pushed. 'They told me she was found at a distance of 9 metres from the cliff, and it's a vertical drop of 30 metres,' he says. 'The launch speed would have to be 3.5 m/s, which is faster than a walk but not as fast as a really fast run. I told them it was too fast for a push.'

Six years later police contacted Cross again, telling him that Byrne had in fact been found 12 metres from the cliff, requiring a launch speed of 4.5 m/s. This seemed almost impossible. Cross and 12 policewomen used a swimming pool to prove that the speed was impossible for a head-first dive with a run-up limited to the short distance from the safety fence. However, an ordinary push would be even slower.

Cross initially gave himself a 1% chance of working out what had happened to Byrne, which was 1% more than the police gave him. They only approached him to rule out a few of their 20 theories. Cross easily ruled out 15 of these, and eventually concluded that Byrne had been 'spear-thrown' over the edge by a strong man. This testimony helped convict her boyfriend, Gordon Wood.

In Cross's 2009 book on the trial, *Evidence for Murder*, he argues that the use of barristers to cross-examine expert witnesses before juries makes no sense. Instead he proposes getting multiple experts together in front of a judge to debate each other's conclusions, as is done for some civil trials. He believes this process, a little like the peer-reviewing that occurs when scientific results are published, would focus more directly on the truth of the matter and increase the chance of justice being done.

Are nanoparticles safe?

Nanoscience, the science of the very small, is an exciting area of research, but there are concerns that nanoparticles may cause cancer and have unpredictable environmental effects. Dr Amanda Barnard's work is considering 'not just what nanoparticles do to the environment but what the environment does to them'.

While the toxicity of nanoparticles may be tested in the lab, Barnard says 'Once they leave the lab they can be changed by the environments they find themselves in, and toxicology tests may be invalid. It's an enormous problem. There are so many combinations and environmental conditions.

'Testing them all in experiments would take many lifetimes, but with a supercomputer you can consider every possibility rapidly. You can also look at high temperatures or strong magnetic fields that might be hard to test in the lab.'

Barnard looks for phase transformations (changes in the structure of crystals that may alter an object's chemical behaviour) that may take place in a nanoparticle and indicate that previous testing may not be adequate. The combination of specific particles with certain physical or chemical conditions can then be lab-tested to see if Barnard's warning is a false alarm.

Computer models may not be perfect, but Barnard says, 'I validated my research on nanogold'. When gold is reduced to crystals 1–5 nm across it shows remarkable properties that are very unlike those of 'normal' gold. 'In every case where one-to-one comparisons were done with experiments, the computer predictions were basically right. That doesn't mean the model will be perfect in every case, but we have to weigh benefits of having some kind of prediction versus having nothing.'

Barnard's work now spans carbon nanotubes, metals and semiconductors. She has also done some exploration of nanowires, but says that nanoparticles, tiny in all dimensions, form the bulk of her research.

Most of Barnard's work goes into identifying the negatives associated with nanoparticles, but she can also help make them more effective. 'One kind of environment is a synthesis environment,' she says. 'By showing how changing the environment during synthesis changes properties, my work is very useful for tailoring manufacturing.'

Nanodiamonds are another area of positive research. These tiny pieces of carbon are non-toxic and offer huge potential for the delivery of chemotherapy. 'Unlike a lot of other nanoparticles, they clear the body within seven days once they have delivered their payload, so they don't overload the organs,' Barnard says. 'They also clump together so they tend to release the drugs slowly over months.'

This slow release is much better for the body than the swift pulse of toxicity associated with current chemotherapy. 'Twenty times less drug is needed when nanodiamonds are used,' Barnard says. 'I'm part of a global team looking at making nanodiamond-based chemotherapy patches.'

Barnard has also worked on titanium dioxide, the nanomaterial widely used in sunscreens and self-cleaning surfaces. There are concerns that sunscreens containing nanoparticles of titanium dioxide or zinc oxide could contribute to cancer, but Barnard says she does not yet have the results to confirm or reject this hypothesis.

A common concern is that nanomaterials are coming onto the market so quickly that regulations can't keep up. 'We already have reasonable regulations for drugs, as well as for ultrafine particles and aerosols, which is really what nanoparticles released into the environment are,' Barnard says. 'We need to make sure that deliberately released particles are funnelled through these regulation systems before we know whether we need entirely new systems.'

Barnard says she 'didn't know what I wanted to be when I grew up' as a child, and was no more interested in science than anything else. No particular revelation led her to undergraduate physics at RMIT University in Melbourne, but rather a series of small steps of becoming interested in one thing that then led to another. From there she moved to a PhD in condensed matter physics at the same university followed by 10 years overseas, first in Chicago and then Oxford. 'When I was at university I felt like everything had been discovered. Nanoparticles seemed like an unexplored field. I'm exclusively computational and theoretical. I'm not good in the lab. Once I started looking at shape and structure I realised my techniques could be useful for environmental effects.'

Although Barnard's background is in physics, she's working at the point where physics and chemistry meet. 'A lot of the papers I read are published in chemical journals.' She says she appreciates the opportunity to work with chemists who are taking on the same problems, 'but coming at it from a different angle'.

Her return to work at CSIRO came about because her husband landed a job in Australia. 'We take turns causing the other to move,' she says. 'He quit his job when I got the position at Oxford.' Barnard has racked up an impressive set of prizes, including the L'Oréal Australia Award 'For Women in Science', the Future Summit Leadership Award and the international Young Scientist in Computational Physics Award. She won the Mercedes-Benz Australian Environmental Research Award just before being interviewed for this book, but topped that a couple of months later by winning the Malcolm McIntosh Prize for Physical Scientist of 2009, Australia's most prestigious prize for scientists in the non-biological sciences within 10 years of their doctorate. She has since added the Australian Academy of Science's Frederick White Prize.

Science communicators

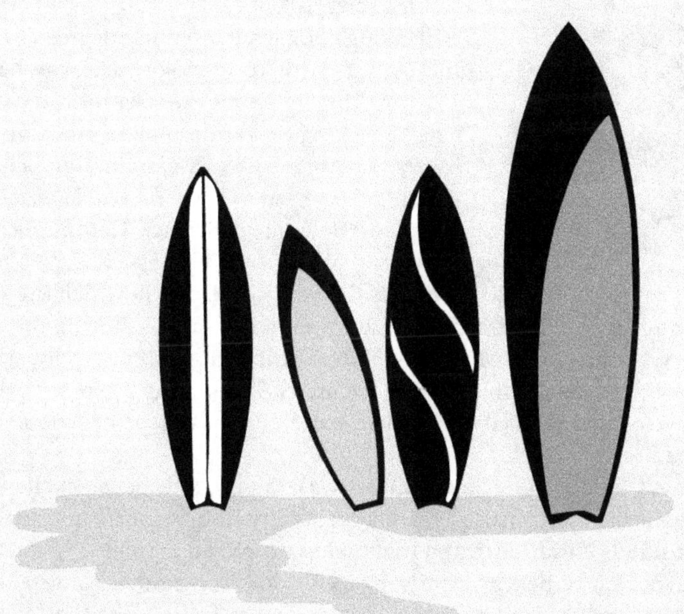

The surfing scientist

Ruben Meerman isn't a research scientist anymore, but he's a demonstration that other exciting jobs exist for scientists.

Having completed a physics degree at Queensland University of Technology, Meerman began his career making optics for the laser industry at Laserdyne Technologies. While applying thin film interference coatings was a fascinating challenge, Meerman was keen to learn more. He enrolled in a Graduate Diploma in Science Communication at the Australian National University.

One component of this course is the Science Circus, in which the students roam rural Australia performing science shows in schools during the day and for whoever wants to come at night. After completing the course, Meerman set up his own science communication business, The Surfing Scientist. Since then he has performed thousands of science shows at hundreds of schools around Australia.

He says the thing he enjoys most is 'seeing the kids get so excited'. 'Children are the best scientists on the planet because they're infinitely curious and have boundless enthusiasm and that makes my job an absolute joy.' Meerman's other favourite audience are disengaged adolescents struggling to connect their schooling with the real world. 'I remember feeling like that so seeing a teenager's expression change from indifference to complete fascination is one of the most rewarding things I can imagine.' Meerman has also taught science education at Griffith University to primary teaching students.

Meerman's rise to stardom started when he proposed a project to teach forensic science in school and convinced Griffith to employ him to run it. A 1999 pilot ran in the Gold Coast region, and two years later it went national. More than 130 000 students from Year 5 to Year 12 tried to solve the question 'Who stole the Minister's Malibu?'

To increase student interest, both the victim and the suspects were famous. Then Science Minister Peter McGauran volunteered to be the first national year's victim, and suspects included JJJ's Adam Spencer, rock stars, and a professional surfer. Meerman then took the concept to the UK, involving half a million students.

Students conducted experiments using forensic kits containing 'crime scene evidence' supplied by Meerman. As well as providing students with an opportunity to practise science with a connection to the real world, the program gave JJJ and ABC Online opportunities to run interviews and publicise facts and case studies relating to forensic science.

The success of these programs made Meerman in demand as a TV presenter, and he now regularly appears on *Catalyst*, *Rollercoaster* and *Sleek Geeks* as well as continuing to perform science demonstrations at corporate and community events and schools. His series of books of fun science experiments or 'tricks' are a spin-off from his ABC website, also titled The Surfing Scientist, where he publishes lesson plans, teacher demonstrations and science conundrums.

At one point Meerman enrolled in a PhD researching shark nets, but the research was put on hold as the communication career took off, and the project has yet to restart.

Meerman was no science enthusiast at school. While others were saving up to buy telescopes or filing collections of interesting rocks, he says 'I was surfing'. He only got into science because 'the best-looking girl in the school was doing physics, and I wanted to be in the same class as her'. However, so many other people signed up for physics that year the school created an extra class, and Meerman ended up in the wrong one.

Despite not doing particularly well initially, Meerman fell in love with physics and majored in the subject at university. He was also fascinated that merely telling people he was studying physics elicited responses like 'I didn't know you were a brain'. 'I'm of very average intelligence and definitely not a "brain" but it shows that physics was and still is perceived as being much more difficult than it really is.' He admits that being thought of as 'smart' might have influenced his decision to stick with science, but his passion for practical, hands-on science did eventually result in some reasonable marks. 'I loved the experimental subjects in my degree and scored well enough in those to get me through university. My first job was almost entirely hands-on and involved powerful lasers, big vacuum chambers and 10 000 volt electron guns which satisfied my thirst for gadgets and danger.'

'I fell in love with physics because reality is almost always much more surprising and bizarre than you'd think,' Meerman says. 'When I started surfing, I used to think waves were big blobs of water travelling across the ocean. Then I discovered that in deep water, the water molecules don't get relocated at all. They move in a perfect circular orbit and end up where they started when a wave passes by. It's energy that's travelling out there, not water, and that's a beautiful thing to know when you're waiting for that wave!' Experience with water waves has also proven useful in getting to grips with the large parts of physics that involve studying wave behaviour in different environments.

Science communication was not something he had planned on either, but as soon as Meerman heard about the Science Circus he knew he wanted to be in it. However, if his path to both science and science communication was accidental, he now has no doubts that this is where he wants to be. His page on ABC Online announces 'I reckon I've got the best job in the world.'

'Science keeps progressing at an astonishing rate so I hope to keep doing this for the rest of my life because there will always be amazing new stories to tell,' Meerman says. 'Plus I have no doubt that every new generation of kids will love watching and doing science experiments so I'm just stoked to have found this amazing occupation.'

Taking science to the media

If the quality of reporting of science-related issues in Australia is improving, one person to thank is Dr Susannah Eliott. She is the founding CEO of the Australian Science Media Centre, and her work is as important to the health of Australian science as that of any researcher.

The Centre is inspired by the British Science Media Centre, with both having been established by brain scientist Baroness Susan Greenfield after the 2000 House of Lords report on Science and Society. The report found there was an urgent need to improve communication between scientists and the general public, including raising the quality of science reporting in the mainstream media.

'We target the news media because specialist science reporters generally do a very good job,' Eliott says. 'However, when you have someone who is reporting on sport yesterday and avian flu today they understandably struggle.' The Centre responds to media inquiries by finding experts who can explain particular topics, but it also takes the initiative to provide information about topics it considers important.

In the Centre's second capacity, Eliott is in charge of events such as briefing sessions where a range of scientists present their views and explain their research relating to a particular hot topic. The Centre also does 'round-ups' providing short comments from people with relevant expertise.

The Centre has only been operating since late 2005, and almost immediately had success stories. Eliott gives the example of a briefing on the 'abortion drug' RU486 held the day before the Senate voted to pass control of the drug to the Therapeutic Goods Administration. 'Journalists said it was the first time they had heard real facts,' Eliott says, despite the issue leading the news for weeks beforehand.

'There was so much misinformation about, it was good to correct some of the bad science.' Eliott says. In that case the issue was all over the national agenda, and journalists were hungry for the facts. At other times the Centre can draw attention to something that might otherwise go unreported. 'Our first briefing was on the World Meteorological Organisation's annual extreme weather report. It's something that is usually missed. There were 110 items in the media, including some overseas. Normally there are three or four.' Other issues informed by the Centre include bushfires, climate change, nuclear energy, stem cell research, cancer, and carbon trading.

The Centre's work is not always controversial. In 2009 the Centre was contacted by a couple of whale researchers who had just witnessed a baby whale take its first breath. The images sent out by the Centre captivated the attention of the Australian and international media. 'It was nice to give an inspiring story a nudge and help out a couple of researchers who had no other avenue into the media,' Eliott says.

Where there is more passion, for example on genetically modified foods or embryonic stem cells, Eliott says that the Centre is aware that it may be accused of bias in its selection of scientists. She says, 'People need to understand we are not representing a weighting of the scientific community.' Rather, the Centre provides a range of scientific perspectives for the media to use. 'It's important in the long run we are not favouring one viewpoint,' she adds, but every briefing cannot be perfectly balanced.

One danger for an organisation like the Centre is that it might be pressured not to publicise bad news for its sponsors. Governments don't always appreciate scientists telling the world their policies are damaging the environment or putting lives at risk, for example, even if the scientists have the research to support these claims.

To avoid being subjected to excessive influence, the Centre caps its funding from any particular source at 10% of its budget. Many of the sponsors are news agencies with a vested interest in getting accurate science coverage. Other sponsors include CSIRO, Shell Australia, and the Cochlear Foundation.

The science advisory panel includes Professor Patricia Vickers-Rich (see page 2), along with other eminent names like Sir Gustav Nossal and Nobel Prize winner Professor Peter Doherty.

Eliott was headhunted for her position, having previously spent six years in Sweden as Communications Manager and then Deputy Director at the Geosphere/Biosphere program, a network of 10 000 scientists in 80 countries studying the Earth's life support systems.

As a child, Eliott was 'really interested in science, but it was a kind of rebellion'. Her parents were artists who believed school was something to grin and bear only for as long as you had to. Eliott enrolled in the Sydney College of the Arts to do painting, but a simultaneous enrolment in science at Macquarie University won out. Her PhD at Macquarie examined slime moulds, collections of single-celled organisms that can behave as a single organism and even show remarkable decision-making abilities.

The shift to science communication came after doing a course under Dr Peter Pockley at the University of Technology, Sydney, in the Centre for Science Communication. Pockley later sent Eliott an advertisement for a job there. While she juggled research with teaching scientists how to communicate their work for a while, she gave up the moulds when she had a child.

'It wasn't an easy decision to give up research,' Eliott says, 'but science [research] is a tough row to hoe with all the grant proposals you need to submit.' Her husband works as a research scientist. Since making the jump to communication Eliott helped establish 'Science in the Pub' and 'Science in the Bush', and acquired a Graduate Diploma in Journalism.

Fires, killer whales and megafauna

Dr Danielle Clode's career is an inspiration to anyone who loves science, but doesn't want to be tied down to one particular topic. With an undergraduate degree in arts and a PhD in zoology she now combines scientific research across many fields with a career as a science writer.

At the University of Adelaide, Clode majored in animal behaviour as an area of psychology. On winning a Rhodes Scholarship she found that Oxford taught her specialty as part of zoology. Her doctoral thesis examined the impact of feral mink on the behaviour of seabirds in the Outer Hebrides.

Clode found that the mink have had a significant impact on the ecology of the islands since being introduced. 'The birds have moved to the offshore islands, and are also nesting in larger colonies, which is probably a response to the mink,' she says. Mink represent a particular threat to the birds because they are both aquatic and terrestrial, enabling them to attack nests in places that non-swimming animals cannot reach.

Since finishing her PhD, Clode has been an associate at the University of Melbourne, and has conducted some zoological research. However, most of her time has either been taken up as a science writer or doing short-term contracts on apparently unrelated projects.

The most recent of these has been a study for the Country Fire Authority of the psychology of those affected by bushfires, including how local fireguard groups responded to the 2009 Black Saturday fires. This dovetails nicely with her latest book, *A Future in Flames*, a look at fire in the Australian landscape, and what we are facing in a changing climate.

Clode's previous book covered very different territory: the behaviour of Australia's extinct megafauna, the giant marsupials and birds that roamed the continent until around 50 000 years ago, including wombats the size of rhinos and 3-metre high kangaroos. 'I modelled it on a field guide, including maps of distributions as if they lived here now,' she says. She also tried to 'feel how the animals fitted into the landscape'. The book was commissioned by the Museum of Victoria but the potential topic was so big that Clode restricted herself to the species that survived into the late Pleistocene era, shortly before the great wave of megafauna extinctions.

The work took her back to her favourite undergraduate subject. Theoretically the topic was animal behaviour in general, but her original lecturer, the late Frank Dalziel, had focused on the behaviour of extinct animals – a difficult area in which to verify one's theories. Clode explains that Dalziel had taught about the behaviour of whales and dolphins, only to find himself deluged by students wanting to study the animals in the field.

The university's budget didn't stretch that far 'so he shifted to dinosaurs because you can't do projects on them,' she laughs. Whatever the motivation, Clode notes that documentaries like *Walking with Dinosaurs* depend on just this sort of work.

'It's not a huge area of research, but it is a fascinating one.' She adds that artists who draw extinct animals rely on theories of behaviour as well as fossil anatomy.

Clode has published papers on the role of prescription drugs in heroin and methadone overdose deaths. She has also been published in respected peer-reviewed journals on the psychology of posing for photographs. 'I do a lot of writing and editing for medical and environmental organisations,' Clode says. 'Sometimes that blurs into actual research. There's a lot of demand for research skills, and it can be hard to find the right person, so if you're there you can be snapped up.'

Another of Clode's books, *Killers in Eden*, considers the remarkable history of whaling in Eden, New South Wales, where for almost 100 years killer whales worked cooperatively with human whalers, herding baleen whales into Twofold Bay for humans to kill. 'I wanted to consider the story of cooperative hunting from the killer whales' perspective,' Clode says. She used the book to consider whether the cooperation predated European arrival, and whether the Eden situation was unique. The book is the basis of an ABC documentary.

Along with a history of the early French naturalists' studies of Australia's flora and fauna, *Voyages to the South Seas*, Clode has published on the history of the Victorian Land Conservation and Environment Councils and written *Continent of Curiosities*, a study of 11 local artefacts held by the Museum of Victoria. 'Each essay is on a particular exhibit, but I try to show the interconnectedness of science and how a line of investigation can take you unexpected places,' she says. It's a story she's well placed to tell.

Science-ology

Science and comedy are not words that usually accompany each other. However, there's a growing field of comedians making science-based jokes. One of them, Ben McKenzie, has achieved success with his science-based humour at comedy festivals, during science week and through museum tours.

One successful recent venture is Museum Comedy, which started out in 2008 with McKenzie and a team of comedians leading audiences around Melbourne Museum, providing a somewhat offbeat take on the exhibits. Scientific authenticity is carefully respected, though the science-to-joke ratio varies between comedians. The concept has done wonders for shaking the image of museums as worthy but dull institutions, selling out in Melbourne in 2009 and expanding to Sydney in 2010.

One of McKenzie's stand-up shows was built around an analogy between science and rock'n'roll. McKenzie suggests that science is naturally wild and rebellious like the best rock music. However, just as record companies try to tame rock and turn it into homogenised pop, McKenzie believes governments, corporate financers and the media are inclined to undermine what makes science challenging and interesting.

He calls for a new movement, Science-ology, to break the bonds and let science be science. McKenzie quickly distinguishes his movement from its near namesake, Scientology, with a quick tour of the history and beliefs of the controversial religion behind the free personality tests.

Making fun of a creed that holds that trillions of years ago the Earth was occupied by an alien dictatorship that left the world riddled with unattached souls responsible for all our psychological hang-ups may be a bit like shooting ducks in a barrel, but McKenzie quickly moves on to tougher game.

He considers the way science is distorted in television adverts and the wider media, particularly the use of Sam Neill to spruik the supposed evolutionary necessity of meat eating for brain growth. McKenzie points out that Neill is using the credibility he has gained from playing scientists in films to push a scientific claim for which there is actually very little evidence.

On the way he gets in plenty of jokes about *Jurassic Park* and less successful Neill films – 'And if you haven't seen *Event Horizon* … I envy you.'

McKenzie also covers attempts by governments to silence scientists trying to discuss their work on controversial topics such as global warming, and the shortage of television shows that make science interesting. The show ends with a re-enactment of Dr Julius Sumner Miller's famous performance getting an egg to drop into a bottle by lighting a fire under it (somewhat oddly used to advertise chocolate in the '70s and early '80s). McKenzie isn't just interested in entertaining – he tries to make sure the audience actually learns something about the topics, including the physics of the demonstration.

McKenzie's previous show was more specific, looking at the development of life on Earth and the history of evolution. That show ended with McKenzie saying that the aim of all life was to live as long as possible and to have sex, so he encouraged the audience members to do just that. He was taken aback when performing in a bar in Adelaide and one audience member belligerently demanded to know why he should follow McKenzie's advice.

However, McKenzie doubts hecklers will ever top a Scientologist who kept interrupting him as he discussed the Church's beliefs. 'I told her I was going to do the show anyway because it was the only thing I had written, but she could tell me I was wrong afterwards. Perhaps unwisely I gave her my email address and she sent me lots of information which was better than anything I had before, so I used it in the later shows.'

More recently, McKenzie celebrated science weeks by summarising Stephen Hawking's *Brief History of Time* in an hour, and producing a show about dinosaurs for children. He says, 'Dinosaurs are to 10-year-olds what girls or boys are to 15-year-olds: they think about them all the time, they have pictures of them all over their walls, and they'd love to touch one but they're afraid of them all the same.' Outside comedy he has presented educational DVDs, mostly on science topics.

Critics have been generally positive and McKenzie says audience response has also been good. 'Sometimes you get audiences who laugh uproariously. Others don't laugh much but applaud a lot at the end. I think they're interested in the science as well as the gags.'

McKenzie studied science at the University of Melbourne, but never completed his degree. 'I wanted to be an actor but thought I needed something to fall back on. I was studying a lot of computer science and even though I still do that as my day job I had worked out I didn't want to be a computer scientist,' he says. 'It was like at school where I wanted to do a bit of physics and a bit of chemistry and a bit of biology and was told you couldn't do that.' At the same time McKenzie was heavily involved in student theatre, particularly the Melbourne University Comedy Review.

During a dry patch for acting roles, McKenzie realised, 'If you want to be a performer you need to make your own work' and decided to create his first show. 'I like comedy which is about something the comedian is passionate about, not about how your mother-in-law annoys you,' he says. Science seemed the obvious choice, although he says he made the mistake in the first show of being apologetic for being so keen on 'something so geeky'.

McKenzie says he is a voracious reader of popular science books and often thinks, 'That would be good in the show', but he is also careful to research topics. 'I want to make sure that I'm conveying real science.'

Audiences at the first shows seemed to be mostly science students and teachers, but McKenzie says he is now getting a more general audience, 'Which is good, because those are the people I really want to reach.'

Zoologists

Do kangaroos have friends?

Winning first prize for research among postgraduates at a major university is a great way to start your scientific career. Doing it when still an undergraduate is extraordinary, but for Alecia Carter that was only the beginning. Within a few years of finishing her undergraduate degree she's already broken ground in our understanding of the personalities and social networks of kangaroos and baboons, along with gaining insights into several other animal species.

Conferences give researchers a chance to inform colleagues of their latest work, usually using large posters reporting their results. Postgraduate students at the University of Queensland are given a similar opportunity, with prizes awarded for impressive and innovative research.

Carter's winning poster was based on her research on the mating habits of eastern mosquito fish, an introduced pest that is displacing native fish from swampy waters. In this species, the males mate by racing up behind females, and Carter says reproductive success 'is all [about] the males' prowess at swimming'. However, the low levels of oxygen in swampy waters mean that the fish are forced to spend 'a lot of time breathing at the surface. Presumably this takes time away from chasing females and exposes them to predators, which is bad.' Carter set out to establish if the fish could compensate.

The paper resulting from Carter's work was published in the *Journal of Experimental Biology*, the most prestigious journal in the field, another exceptional achievement for an undergrad.

Carter found that after six weeks acclimatising to a hypoxic (low oxygen) environment the males had adapted and were able to mate frequently, whereas newly introduced males could not. Nevertheless, the acclimatised males had unusually low sperm counts, proving there's still a price to pay for living without much oxygen.

The project began as a 20–30 hour experiment, part of a program to give undergrads a taste for research, but with encouragement from supervisor Dr Robbie Wilson it ended up taking four months of Carter's free time.

Having outshone students years ahead of her, Carter moved on to female grey kangaroos for her Honours thesis.

Like many other mammals, grey kangaroos have social networks that come and go, with animals inhabiting overlapping ranges, a social organisation called fission–fusion. Little research has been done on such social structures, and Carter believes marsupials are particularly under-studied. 'People regard marsupials as less than true mammals. It's one of the things I'm hoping to change,' she says. 'We have the coolest wildlife in the world, but we under-appreciate it.'

Carter set out to discover whether female greys have friendship networks, measured by whether they spent more time with some individuals than others. She found the answer was yes.

Delving further, Carter investigated whether an individual seemed to get any benefits from time spent with her friends. She found that when among friends a kangaroo would spend more time feeding, and less watching for predators, than she would when among strangers.

Carter also compared a kangaroo's sociability with other aspects of her personality. She measured each animal's boldness both by seeing how closely she could approach them before they would take off, and by introducing something they might fear. 'My original idea was to use a stuffed dog, but that would have been difficult to suddenly uncover in the field,' says Carter. 'Then I came up with the idea of a stuffed python.' One of Carter's supervisors suggested a live python might be more effective. 'Well, I'll see if I can get ethics committee approval,' Carter replied.[1]

For most people, actually acquiring a 2-metre long python might be as much of a challenge as getting the idea past an ethics committee. Fortunately, Carter happened to have one lying around, as it were, since she and her boyfriend keep snakes as pets. 'For us a typical date is going up to the nearest national park and looking for snakes,' she says.

The python proved a great success, scaring some kangaroos while others were brave enough to come up and sniff it. However, Carter found no correlation between courage in the face of snakes and sociability with other kangaroos.

Carter hopes her work will encourage people to recognise that kangaroos have personalities, and become more inclined to protect them. She says it will also be helpful for programs keeping animals in captivity. 'You can't assume that if you put two animals together they will get along as they would if they had the choice of who to associate with.'

---

1 Research that might cause harm or distress to humans or animals needs to satisfy an ethics committee. For research to receive approval, the ethics committee must conclude that the benefits from what will be learnt outweigh the risk of harm or distress.

The research fulfils Carter's childhood dreams. She says, 'All I ever wanted to do was look at cool animals. When I was six I said I wanted to be a vet because that was the only job I knew working with animals, but when I did work experience there I discovered it was boring – you did the same thing each day.'

'I thought, "this isn't like David Attenborough documentaries", and then I discovered you could actually do animal research and get paid for it.' She has spent every opportunity volunteering on research, with target species including black sea turtles, satin bowerbirds, spot-tailed quolls and various endangered snakes.

Carter's work won her the 2007 Undergraduate Studies (Science) Queensland Smart Women – Smart State Award. She used the $2500 to help fund a six-month trip to Africa to study mongooses (she prefers 'mongeese') and the diverse personalities of Namibian rock agamas, a lizard that looks like it's taken a bath in a paint pot. She's hoping her paper on agamas will be published soon. To raise the remaining costs she did lots of exam marking. 'I'm dreading going back to waitressing; there aren't a lot of jobs in science when you only have Honours,' she said at the time.

Once back from Africa, Carter began to rectify the situation by starting a doctorate at the Australian National University, studying baboons. Carter describes this as 'similar stuff as with the kangaroos but on a grander scale and also looking at how different personality types deal with acute stress.'

Carter's enthusiasm for her subjects is irrepressible. She considers baboons 'by far the most entertaining animals on the planet and I love them and want everyone else to love them.' She describes a seven-month field stint as 'really, really difficult, but incredibly rewarding. We have to be at the baboon's sleeping cliffs before dawn, which means getting up at 4:30 every day, following the baboons for the entire day doing focal observations wherever they go (which is usually up and down cliffs!) until they go to another sleeping cliff when it gets dark again and we know where to find them in the morning (they don't move at night). So days are over 13 hours long, up to 15 hours in the summer.'

Some waitressing was required to save the money to get started on the PhD, but Carter says the baboons 'are so great, though, it's totally all worth it!'

Along with her Bachelor of Science at the University of Queensland, Carter completed a Bachelor of Arts in Spanish and Indonesian. 'I like languages and would really love to work in those countries and it's important to at least try to know the language,' she says.

Wherever her future research takes her, Carter seems to be on target when she says, 'I aim to have an interesting life.'

Refugee solves Australian problems

In the hottest months of 2003, a ship carrying thousands of sheep for the live export trade was rejected by Saudi Arabia. Unable to return to Australia for quarantine reasons, the *Como Express* sailed aimlessly around the Indian Ocean while its cargo suffered and embarrassment grew.

The tale of the dying sheep was big news for weeks, but even among those who remember the events few realise it was a scientist who solved this crisis. Fewer still would realise that Dr Berhan Ahmed was a refugee who has produced work of even more long-term value to Australia.

Initially at CSIRO and now at the University of Melbourne's School of Land and Environment, Ahmed has been seeking sustainable solutions to the problem of termites, estimated to cost Australia AUD$910 million per year in damage, control and repair.

One of Ahmed's research interests was on the use of crushed rock beneath a house's concrete slab to create an impassable barrier to termites. He and his colleagues confirmed that Australian granites need to be crushed to a size that has previously been shown appropriate for American basalt. 'The grains must be between 1.7 and 2.4 mm in diameter,' Ahmed explains. 'If they are smaller than that the termites can carry them. If they are larger there will be holes a termite can squeeze through.'

Termites' main entry points into concrete slab buildings are through cracks, around perimeter edges, and via entry holes around drainpipes. 'To us the gap around the pipe is very small, but to a termite it's a freeway,' Ahmed says. Suitably crushed rocks are now on the market as a result of this work by Ahmed and his colleagues at CSIRO.

In an effort to find other solutions to the termite problem in the Northern Territory, Ahmed has established 80 structures that he refers to as 'dog houses' (others have called them an 'all you can eat buffet' for termites). A variety of treatments have been used on the houses to see which are the most successful at keeping the termites out. The land is leased from the local Aboriginal people, and Ahmed has been working with the Yolngu community to see how

Indigenous knowledge can be used to help us 'learn to live with termites'. He says, 'It is exciting to learn from the locals of Arnhem Land who have coexisted with termites for so long.'

Ahmed believes that it is essential that we adapt to termites, because the old methods of spraying are not sustainable. 'The organochloride sprays are carcinogenic, undiscriminating, and not biodegradable.' He even thinks termites can be of use to us. 'They break down cellulose and they are a good source of protein for chickens, lizards, snakes and fish,' he says. 'The by-products are organic materials, which are nutrients.' Of more than 300 species of termite in Australia, only about 20 cause economic damage.

Ahmed's current research is developing an alarm system for buildings and power poles using a remote sensing system to detect termites before they damage properties and other timber. One of his PhD students is studying DNA 'fingerprinting' of termites for identification and mapping.

Ahmed was born in what is now Eritrea, but fled when he was 15. He 'was lucky enough' to have been able to complete his schooling at a United Nations school in Sudan, where he won a scholarship to study plant protection at university in Egypt. With recognition as a refugee, he was offered the chance to come to Australia or North America, and arrived in Melbourne when he was 25.

'I started out as a tram conductor to learn the culture, practise English language skills and establish friendship with mainstream Australians, and in the hope of getting reference and support letters for employment and rental accommodation.' After 10 months he went to La Trobe to study animal science. He recalls, 'I was offered a job at CSIRO, which was an opportunity I could not pass up.' While at CSIRO he completed a PhD at the University of Melbourne, and after working at both institutions eventually settled at the university.

Ahmed says he was always interested in agriculture and forestry. The more he learnt, 'the more interested I was'. When the *Como Express* became news, Ahmed realised there was not likely to be a quick solution. 'I knew that you could not solve the problem without knowing the culture of the area. People were trying to intervene without knowing the culture.'

Ahmed called authorities who were trying to find a destination for the cargo, and suggested they should donate the sheep to Eritrea rather than searching in vain for a buyer. At first there was little interest, but as hopes for selling the sheep waned the idea became more attractive.

However, another problem arose. The Eritrean government was dubious about accepting sheep no one else wanted, despite the extensive hunger there at the time. Ahmed was able to use his credentials as an animal scientist to assure contacts among Eritrean emigrants that the sheep were safe.

'I said, "From the science I can tell you there is no problem, but from the politics and image I can't resolve".' However, as dialogue grew between the Australian and Eritrean governments, the outstanding issues were eventually resolved. Some Eritreans remained critical, seeing the acceptance of sheep others had rejected as humiliating, but most people saw this as a win–win solution.

This was not Ahmed's first venture outside scientific research. He was active in Eritrean organisations that opposed Ethiopian control of his country before it gained independence, and more recently he has worked against the current Eritrean dictatorship. He brought a professor specialising in Horn of Africa studies to Australia to explain some of the region's issues to relevant agencies.

Locally Ahmed has drawn together groups representing refugees from various African communities to promote education, cultural exchange and better understanding between new immigrants and the police at Housing Commission estates where tensions have sometimes run high. Ahmed initiated and established an African Think Tank organisation with colleagues from the various African communities to promote and advocate on issues affecting refugee communities in Australia.

Ahmed has run as a candidate for the Australian Greens at the 2002 and 2004 state and federal elections.

On Australia Day in 2009, Dr Berhan Ahmed, the Chairman of the African Think Tank, went to Canberra to attend the ceremony of presentation of Australian of the Year. As the Victorian Australian of the Year 2009, Ahmed was one of the nominees for Australian of the Year. He describes it as 'A long, long way from when I arrived here twenty-one years ago, a scared refugee coming from terrible conditions as refugee in Sudan and then in Egypt under the auspices of United Nation High Commission for Refugees.'

The real Batman

The rainforests of northern Queensland are home to a unique research station. As part of his efforts to restore and study this ecosystem, Dr Hugh Spencer uses a miniaturised automatic radio-tracking device he invented for thumb-sized bats. Spencer, a biologist and engineer, was awarded the 2002 Science Communicators's 'Unsung Hero of Australian Science' award for establishing and managing the station and for the technologies he has invented to help study bats.

After a fairly erratic undergraduate career (starting at the University of New England and eventually graduating from the Australian National University in 1965), Spencer started a PhD at Monash tracking echidnas, shifted to physiology (and electronics), was awarded a scholarship to the University of Manitoba where he did a MSc and PhD on neural transmitters in the rat striatum[1] (and was converted to a conservationist). After a post doc working on the hippocampus at the University of California, Irvine, he returned to Australia to teach neurophysiology and botany at the University of Wollongong, where he got what he calls the 'rainforest bug'.

Spencer demonstrated his commitment to rainforest research when he and his wife Brigitta sold their Wollongong house and cashed in their entire savings to buy an 8.5 hectare strip of cleared land running from the highlands to the coast at Cape Tribulation in 1988. Two decades later, most of the property has been reforested and they have established one of Australia's very few independent terrestrial research stations – the Cape Tribulation Tropical Research Station – in the seasonally wet tropical rainforests of the coastal lowlands.

The volunteers who do so much of the work at the Station are mostly biology and environmental management students coming from as far away as Alaska and Turkey. Besides helping with forest restoration, cooking and cleaning, these students take part in research projects relating to the rainforest, many bat-related, and are taught the essentials of low-impact renewable-energy based living. The Station also hosts researchers, students and interns who

1 Part of the forebrain involved in reward and motivation.

carry out a wide variety of research projects on subjects as diverse as flying-fox urine, coconut invasion, and low-energy building technology.

A visitors' centre – known as the Bat House because there is always at least one orphaned flying fox in attendance – helps to fund the station and educate tourists. Mega-bats (flying foxes and their relatives) have an essential role in rainforests as pollinators and spreaders of seed, and are a vulnerable part of the ecosystem.

Besides the financial cost, Spencer has had to put up with a lot to get the station going. It was founded in the midst of a dispute over World Heritage listing for the area, when many locals were very keen to see the forests logged and settled. Since then there have been plenty of heated debates about the region's future. The Station has taken a staunchly pro-environment position, not always welcomed by local business owners.

'The Daintree lowland rainforests have a pedigree of over 100 million years,' Spencer says. 'It's one of the last remaining remnants of the Gondwanan rainforests.[2] It is incredibly species-rich and worth conserving for its intrinsic value, let alone for its commercial use – ecotourism and bioprospecting[3] plants, animals and fungi for unique compounds – and the fact that it is one area of rainforest in the world that can really be saved.'

The announcement of his unsung hero award noted that Spencer 'has endured personal attacks' and been 'caught up in academic backbiting and turf wars, ongoing financial uncertainty,' not to mention 'mosquitoes, humidity and cyclones!' In interview, he saw little point in discussing past academic disputes, and noted: 'With mosquitoes, you can swat them if you're fast enough, but the malaise that presently affects much of science in Australia is having serious impacts on an independent organisation like ours, especially on funding and collaboration.' He added, 'being independent gives us an enormous amount of freedom to pursue our goals'.

As for 1999's Cyclone Rona, Spencer said it 'did no physical damage to the station, but it knocked over so many trees it took us two days to chainsaw our way to the front road'.

The land had been almost completely cleared in 1970 for cattle. To reforest it, volunteers push down the grasses using the not-so-high-tech method Spencer refers to as 'stomping', and then spray them. A little regeneration comes naturally thereafter, but volunteers plant pioneer trees sourced from the local area. 'Nothing, but nothing, is introduced,' Spencer says. Now over 85% of the area is under dense revegetation.

Spencer stresses that for all their work, the land hardly resembles its natural state. 'It's the old story: you can destroy in seven hours what will take 300 to

2 Gondwana was a supercontinent containing Australia, Africa, South America, Antarctica and India during the age of the dinosaurs.

3 Studying natural products for medical or other applications.

400 years to restore. All we can do is start some degree of healing, which you can get in a year or two. Trees grow real fast here, the 21-year-old forest area is starting to look pretty good, and we've developed some very effective low-cost, low-effort approaches to rainforest rehabilitation.'

Creating a functioning station hasn't stopped Spencer's research. He has published 10 refereed papers since establishing the Station, and delivered many more to conferences. With increasing student loads, finding time to write up his research, let alone run a station, is becoming increasingly difficult. Much of the research has focused on flying fox conservation and their role in the rainforest ecosystem. He is developing techniques for studying flying foxes with GPS tracking devices and radar.

Other papers include one demonstrating the importance of fig wasps to the pollination of cluster fig trees, which are considered a 'keystone' species essential to the wider ecosystem. Dr Spencer was particularly keen for readers to access the station's website (www.austrop.org.au). Among stunningly beautiful images of the station's location and specially designed low-energy, low-impact buildings, the site seeks students and researchers interested in conducting long-term research on local species – including amethystine pythons, a species that grows to 8 metres long. Other potential projects relate to climate change, forest rehabilitation, low-energy building technologies, and much else besides.

Animal intelligence

Dr Emma Collier-Baker has rewritten ideas about canine intelligence, revealing why dogs have apparently been 'passing' tests it was previously thought they shouldn't. Dogs, great apes and two-year-old children were all getting the same results on a particular kind of test, but solving it in different ways.

In the 1930s child psychologist Jean Piaget created the 'invisible displacement test' for the ability to think about hidden objects. The test involved placing a desired object such as food or a toy in a small container and sliding it behind a series of small boxes. The object would then be transferred from the container into one of the boxes. The subjects of the test could not observe the transfer, but would be shown that the container was empty after it returned to its original position.

Two-year-old children are generally able to pass this test, working out where the object was now located, as are great apes. Younger children and most other animals fail.

These results are consistent with most other tests comparing intelligence across species. 'Dogs were a surprising exception,' says Collier-Baker, a postdoctoral psychology researcher at the University of Queensland (UQ), 'passing this test when they fail on many others'.

However, Collier-Baker repeated the tests under more stringent conditions. 'We showed the dogs had effectively been "cheating" to pass the test and were simply going to the box closest to the small container.' When the container was not left next to a particular box the dogs were stumped. It seems dogs had formed a simple association that gave them a moderate level of success in the absence of such controlled conditions. Thirty-five dogs of a variety of breeds were used.

'In contrast, 21 two-year old children, tested with identical apparatus to the dogs, were able to pass the task no matter what the proximity of the small container to the target box,' Collie-Baker says.

Collier-Baker did the same controlled version of the test with two chimpanzees and found they were able to pass the test under all conditions. In contrast, a spider monkey failed the standard task, while gibbons put in a mixed performance, prompting further research.

'Gibbons are lesser apes and these primates have been surprisingly understudied despite their value for reconstructing the evolution of primate cognition[1],' Collier-Baker adds.

Collier-Baker hastens to point out that dogs' ability to achieve success on the standard task through simple associative means, rather than what she calls 'representational capacity', still demonstrates an impressive skill.[2] 'Many other species tested simply failed the task. Dogs found a way around the representational requirements in the traditional task and still found the object much of the time,' she says.

Dog owners love to tell of pets capable of retrieving objects hidden long before, or remembering objects by name. Collier-Baker points out that while studies showing the size of some dogs' vocabularies are 'really interesting' they display a different sort of intelligence from that required to work out where the toy must have been transferred to.

Collier-Baker's work has been rewarded with publication of studies in *The Journal of Comparative Psychology* and *Animal Cognition* and an award from the American Psychological Association for the best paper published in this journal that year.

As a child Collier-Baker hoped to be either a journalist or a vet, and says she has found a career path that incorporates both a passion for writing and a fascination with the natural world.

Collier-Baker grew up in the UK and studied English Literature at Newcastle University. When she was part-way through her degree her family moved to Australia and Collier-Baker transferred to the University of Queensland. The Australian university system provided more scope for elective subjects, and Collier-Baker took to philosophy and psychology with enthusiasm.

While seeking advice about the possibility of doing an Honours project on animal cognition, Collier-Baker heard about the work of Professor Thomas Suddendorf, who was doing a study on dolphin cognition. He supervised both her Honours and PhD projects.

Inadequate study design can be particularly problematic for research on animal cognition, with some subjects beating tests by smelling out objects or using other simple strategies. Collier-Baker's study with dogs highlighted the importance of stringent control conditions.

On the other hand it is important to remove constraints that may return false negatives.[3] For example Collier-Baker recently found that chimpanzees and children are able to pass a 'double invisible displacement test' which they

1 Information processing and applying knowledge.

2 Representational capacity refers to the ability to have words or mental images stand in for physical objects, which is the way in which humans and apes solve the problem.

3 Scientists refer to errors in tests as 'false positives' where the test suggests something is true when it is actually not, and 'false negatives' where the result wrongly indicates something is untrue or fails.

had previously always failed. In a standard double displacement task, boxes are arrayed in a straight line and the reward is invisibly moved via a small container behind two boxes before the container emerges empty. Therefore, the reward could logically be at either of the two boxes, and if subjects choose an empty but visited box they may choose again. This task requires extra thought processing compared to a single invisible displacement test, and it seemed that chimpanzees and young children did not have the capacity to keep two possible hiding locations in mind. When they selected a visited but empty box, subjects tended to then search at the adjacent box inflexibly, as if refusing to believe it could be empty. However, Collier-Baker tried placing the boxes in a novel alignment on a vertical display in which boxes were no longer next to one another, and found that chimpanzees and two-year-old children could pass this equally complex task. This suggests that their problem was that they were too tied to a particular search pattern rather than that they had a problem with representing the hidden object.

Collier-Baker was keen to see if gibbons fill a substantial gap in intelligence between monkeys and great apes. In her work she has found them 'very flighty' and apt to lose interest in tasks. 'Great apes are generally more motivated to sit in front of you for longer periods and are less nervous about novel objects than gibbons,' she says, making them easier test subjects. After visiting zoos around Australia to work with various primates, she spent six months at the Smithsonian Institute in Washington DC on a Queensland–Smithsonian Fellowship. She tested orangutans, gorillas and gibbons on various studies at Think Tank in the Smithsonian National Zoo. One of these experiments found that gibbons cannot recognise the image they see in a mirror as their own. This result was published in the prestigious journal *Proceedings of the Royal Society.* In contrast, great apes are successful on mirror self-recognition tasks.

Collier-Baker returned to Australia and completed her Postdoctoral Fellowship with Sudddendorf, subsequently lecturing third-year developmental psychology at UQ. She recently started an Endeavour Research Fellowship conducting field work that looks at cognition and behaviour in wild Sumatran orangutans in Indonesia.

Collier-Baker's work could help zookeepers make their enclosures and enrichment programs more interesting for primates and she has established a rapport across the species, including with chimpanzees in Rockhampton. When a colleague pretended to scare her with a toy crocodile, one of the chimps rushed over to protect her from the toy and soothe Collier-Baker.

Collier-Baker believes her work will help establish the capacities of humanity's ancestors around the time we diverged from the other apes. She also hopes she can contribute to increasing awareness about the way animals should be treated. 'If I can get people excited about animal intelligence I hope I can help to raise interest in the ethical treatment of animals and in the conservation of our highly endangered great ape cousins.'

The talking ape

Scientist Emma Collier-Baker measures intelligence in the Great Apes (see p. 187) but Dr Carla Litchfield has gone one better. She spent one hot January locked in the primate exhibit at the Adelaide Zoo, seeing the world from an ape's perspective.

Litchfield was not alone in her endeavour. Twenty-four volunteers spent a week each in the enclosure, but she was the only one there for the whole month.

Zoo publicity listed three aims for the project:

- To create awareness of the closeness of humans to their primate cousins
- To provide a platform for research on animal behaviour and enrichment, and
- To raise funds and awareness for the conservation needs of primates in the wild.

The funds and publicity were also a step towards the establishment of a large open-range chimpanzee enclosure, with a temperature-controlled and all mod-cons indoor area, at Monarto Zoo, 80 km from Adelaide.

'A lot of my research is behavioural enrichment[1],' Litchfield says, so 'what better way to find out what it is like' for apes to live in an enclosure than to spend time there herself?

Although publicity suggested the participants would be imitating the apes and living on bananas, this was not really so. Those inside the exhibit could hear what visitors were saying, and Litchfield says a fair amount of discussion and answering of questions occurred.

In some ways the humans did come to resemble the other animals at the zoo. 'We were given a very healthy diet,' Litchfield says. 'Some started begging for junk food from people going past. By the end of the fourth week we had chocolates and drinks coming over the fence. We see apes food-beg. It's a demonstration that visitors should not eat junk food in front of apes.' It also showed how much people struggle to give up sugary snacks and soft drinks.

---

1 Stimulating animals intellectually by giving them more interesting enclosures.

To keep the participants and visitors entertained, each week the residents were assigned a task. These included painting canvasses, putting on a play, or an 'ephemeral fashion show'. South Australian Premier Mike Rann and then federal minister Amanda Vanstone each entered the exhibit for an hour or so to join in the canvas painting, yielding a reward for the zoo when their work was auctioned.

The project also involved a series of role plays which, Litchfield says, 'were the only times they acted like apes. For example there was a mock vet check where the vet would pretend to microchip them or give them implants to prevent them from reproducing.'

One gorilla in the next enclosure watched one event enthralled, sitting on top of a pole for a better view.

The program was considered a success, and something similar may be run in the future.

Litchfield also says she learnt some valuable lessons in regard to caring for the normal occupants of the space. 'You need to monitor temperature variation, particularly in the hot Australian summer. It got to 65°C in some parts of the enclosure.' Remarkably, tests of the temperature had never been done before, so no one knew what we were exposing our nearest relatives to. Discovering how animals adapt to different temperatures in zoos is a flourishing new area of research, with data currently being collected from numerous species housed in zoos in the United States.

'By day three the sense of smell became heightened,' Litchfield observes. 'You could smell cigarettes a mile away and lavender and herbs became overpowering.'

All this despite the fact that the participants were let out at night, because, Litchfield says, 'The gorillas and chimps needed to have their sleep time. We are not the naked ape but the talking ape. Had we stayed in the neighbours would have got no rest.'

While the human ape exhibit makes for good headlines, it is not the only high profile event in Litchfield's distinguished career. She won the Unsung Hero of Australian Science award for 2000, and is on the board of the United Nations' Great Ape Survival Project (GRASP) and President of the Australasian Primate Society.

'As a child,' Litchfield says, 'all I wanted to do was work with animals and be a scientist.' A Bachelor of Science at the University of Adelaide was an obvious first step. The decision to specialise in chimps came after a meeting with Dr Jane Goodall, whose work on wild chimpanzees revolutionised our perceptions of how much we share with the great apes. This led to a PhD in Animal behaviour at Adelaide University studying chimpanzees and caracals, followed by a postdoctoral research fellowship at the University of St Andrews.

Litchfield now lectures in the School of Psychology, Social Work and Social Policy at the University of South Australia when she is not at the zoo.

Having spent a year in the Kibale forest, Uganda, and seeing that a third of the animals were missing a hand or foot as a result of stepping into snares Litchfield decided her life's mission was clear. 'I had to watch them screaming in pain and know there was nothing I could do.'

Litchfield has been back and forth to Uganda many times since. More recently she has led three teams of 12 members of the general public on study and conservation tours to visit wild chimpanzees and gorillas in Uganda, chimpanzees at Gombe in Tanzania, and orangutans in Borneo.

Besides fieldwork she has concentrated on finding ways to enrich the environment for animals in zoos. She is helping ZoosSA develop the area of Conservation Psychology, because the only way to save animals is to change people's unsustainable behaviour. She supervises a number of Honours and PhD students, studying behaviour and cognitive abilities of different species, monitoring human–animal interactions in different environments, and also finding effective ways of changing human behaviour to be more sustainable.

Litchfield encourages those whose interest in the apes has been piqued by her work to give to one of the many groups working to preserve their habitats, but she does not promote a particular organisation. She has also been involved in developing and promoting guidelines for gorilla and chimpanzee tourism since the start of gorilla tourism in Uganda in the mid 1990s. This ensures the visitors whose money helps preserve apes in the wild don't prove their undoing by transmitting disease or damaging the animals' social structure.

Out of this work came *Treading Lightly* (Traveller's Medical and Vaccination Centre 1997), a book she wrote on responsible tourism with the great apes. In particular Litchfield emphasises that travellers in the area should avoid eating any meat whose identity they cannot be certain of, as there is a fair chance that such 'bushmeat' may be from apes or other endangered species. She points out that this practice not only threatens the lives of the other apes, but also probably accounts for the transmission of HIV and ebola to humans.

'People often are unaware of the fact that animals in "the wild" may be suffering terribly,' Litchfield says. 'It is a myth that it is a pristine perfect environment. Some national parks in Africa are completely fenced in, and others are smaller than large open-range zoos, such as Monarto. Scientists have to work with local communities to create healthy landscapes for people and animals in "the wild" and captivity.'

Litchfield has written several books for children from her work, the most recent being *Saving Pandas*. She would like to stress that 'Science is fun and that should be celebrated. Working with animals is fantastic.'

Life in the canopy

The news is good for children who love both science and climbing trees: there is a career that combines the two.

It's called canopy research, and with 40% of life on Earth inhabiting the forest canopy, its importance should not be underestimated. Professor Jonathan Majer, Head of Environmental and Aquatic Sciences at Curtin University, is one of Australia's leading exponents of canopy research. In 2007 found he himself teaching the Canopy Biology Course in Brazil, where postgraduate students were taught both how to make their way up to the tops of mighty rainforest trees and what to do once they are up there.

Majer's interest in science started as a teenager. 'I've always been a collector,' he says, 'and as a young teen I started collecting butterflies. I constructed an ultraviolet light trap to catch moths and it started from there.' After school Majer studied zoology at the University of Bristol, eventually specialising in entomology.

His postgraduate degree was conducted as part of a Study and Serve program at the University of Ghana in West Africa. Majer compares this to the Peace Corps or Australian Volunteers Abroad, with students doing teaching at the university while undertaking research in developing countries.

While in Ghana, Majer was one of the first to use chemical knockdown techniques to study the insect species inhabiting trees. Plastic sheets would be spread underneath a tree, and then chemicals sprayed into the foliage, enabling a complete sample of insects inhabiting the tree to be collected. Majer also researched methods for the control of cocoa pests while he was in Africa. Rather than simply stepping up the pesticide war, he explored the possibility of using ants to eat the pests.

Majer wrote up his work at the University of London before moving to the Western Australian Institute of Technology (now Curtin University).

After coming to Australia, Majer focused on ant research, with much of his work looking at the insects making their way up and down tree trunks. In a study of a jarrah forest he found 1300 species on tree trunks alone, with many limited to small pockets defined by rainfall and perhaps soil type. He is concerned that climate change will affect this diversity.

Majer's return to the canopy came about when a colleague wanted to know why birds seemed to prefer certain eucalyptus trees to others that seemed very similar. Majer revived his knockdown techniques to demonstrate that some trees supported a far wider variety of insect life, and birds were voting with their beaks in favour of the places with more tasty morsels. Since then he has written many papers on insect life in eucalypt trees, including comparisons between east and west coast species, healthy and diseased trees, and trees inhabiting intact forest and those left standing in paddocks.

Majer was invited to Brazil to further his work on cocoa, and has conducted similar work there. 'I looked at trees left behind after deforestation,' he said. 'It was quite encouraging news that they were still full of life even after they had become isolated from other trees.'

Shortly after returning from one trip to Brazil, Majer was contacted by the Global Canopy Program (GPC). Dr Servio Ribeiro of the University of Ouro Preto wanted Majer to help teach his unique canopy course, and the GPC was willing to pay.

The course takes place in the middle of a rainforest reserve. 'For a week we taught them skills in canopy research, and after that they were given mini-projects and had to do the data analysis and produce a paper and PowerPoint presentation,' Majer explains. The students were taught chemical knockdown and other techniques, such as the placement of tree trunk traps to catch crawling insects.

Majer does his research in Australia using cherry pickers, which he says is possible because 'there are lots of trails and the land is firm' as a result of our dry climate. In Brazil, however, getting a cherry picker into the forest would be far more difficult even if there were the funds to hire one. Consequently, about half the training time was taken up with teaching the students how to climb trees with full climbing gear. *Escaladors*, or professional climbers, were brought in to teach this section.

While some may find the idea of combining science with climbing 30-metre high trees enticing, Majer says, 'I'm perfectly happy in the cherry picker, and I was the first one using ropes in the demonstration, but I prefer to supervise from the ground. Actually, I don't like heights.'

Marsupial nutrition

Emeritus Professor Ian Hume is living proof that the slightest chance can lead to a fascinating scientific career. Being an expert on the digestive systems of bandicoots may not win points at parties, but having a special place in the hearts of overseas koalas (or at least their keepers) and an award-winning textbook makes for a lot of satisfaction.

Hume studied agricultural science at the University of Western Australia, and was working at the University of California when his life-changing moment occurred. Chosen to teach a course in wildlife nutrition, he asked the head of department what he should talk about at a seminar for the department. 'There was a long pause,' recalls Hume. 'Finally he said, "Goddammit man, you come from Australia, you're going to teach wildlife nutrition; tell us about the nutrition of kangaroos".' It was assumed this was a topic any Australian animal nutritionist (perhaps any Australian) would be familiar with.

Knowing nothing about the dietary needs of macropods[1] Hume was fortunate to discover the world's only text on kangaroos in the university's library. 'The students all enjoyed the seminar, but I felt like an impostor and vowed that on returning to Australia I'd improve my knowledge.'

First at the University of New England, and then at the University of Sydney's School of Biological Sciences, from which he retired in 2003, Hume built a reputation as Australia's leading expert in marsupial nutrition. Under the circumstances, this status pretty much guarantees him the title of world-leader.

'When I started at New England there wasn't a call for marsupial nutritionists,' Hume notes. 'Scientists didn't understand much about marsupial nutrition. No one said "hey we have to do something about this area", but once we came along we became very useful. Particularly for ecologists and wildlife managers, we now had the basic information to work out what areas provide important food sources for particular animals and whether they can find substitute sources if a feeding area is lost.' A lot of this information is in Hume's 1999 book *Marsupial Nutrition* (Cambridge University Press 1999).

1 Latin for 'large foot', the macropod family includes kangaroos, wallabies, pademelons and quokkas.

Hume notes that there is a huge variation in marsupial adaptability. Bandicoots are particularly flexible: 'They can shift from eating insects to plant roots and fungi overnight if a fire comes through.'

At the other extreme are the notoriously fussy koalas. 'It's true we sometimes get reports of koalas sitting in pine trees eating needles, but the question is how often do they do it,' Hume explains. 'Koalas may nibble on other things, but they can't be the basis for their diet'. For their core diet koalas restrict themselves to a few dozen among the 600 species of eucalypts.

This extreme specialisation evolved because koalas seek 'the right balance between nutrients and toxic components. It costs a lot of energy and nutrients to detoxify leaves with too many phenolics or essential oils.'

Koala fussiness led Hume to invent the koala cookie. Designed to satisfy the most exacting connoisseur, the cookies are made of lumen chaff, water, binding agent, and 10% fresh eucalyptus leaves (or equivalent essential oils). They potentially represent salvation for overseas zookeepers, who otherwise need a eucalypt plantation to keep their charismatic occupants happy.

The koalas were satisfied but, Hume notes, the cookies 'were too much work for zookeepers to put together, so we have gone back to the drawing board, trying to find a simpler alternative.'

Hume's expertise extends to the digestive system of migratory birds. This again occurred by chance, when he met a migratory birds expert at the house of a mutual friend. 'He was telling me about these birds that breed in northern Europe, which is incredibly productive over the summer, and then fly down to Tanzania in the autumn. There is no food for them crossing the Mediterranean, and precious little in the Sahara, so they fatten up before they leave.'

Hume guessed that the birds' guts would shrink during the journey, since the digestive system 'is very expensive to maintain, and heavy for a small bird'. Further study proved him right. 'The system shrinks by 50%, and the small intestine, the most active part of the digestive system, by 65%.' The birds pay a price for this, however. If they spot a Saharan oasis they need to spend at least three days there before their systems recover enough to take advantage of the food on offer.

Hume's memorable fieldwork moments include a night spent shaking trees to catch ring-tail possums. When the possum drops to the ground it is stuffed in a bag. On two occasions this night the possums disappeared before they could be caught. As torches frantically scanned the area, a student exclaimed 'it's on my shoulder'. It seems the possums had mistaken the student for a tree and run up a leg, under an anorak, and onto a handy arm.

Several years ago Hume was studying yellow-footed rock-wallabies in western Queensland. To his astonishment some Earthwatch volunteers assisting him caught a swamp wallaby, 1500 km from its normal coastal territory. It turned out that in wet years the swamp wallabies take to expanding far beyond their normal range. With the drought of the early 2000s, however, Hume says,

'We've been finding them bogged in dams'. Meanwhile the rock-wallabies, adapted not to need to drink free water, survive in arid lands.

Hume has officially retired, but he keeps the title Emeritus Professor and continues to do some research. His retirement has left a shortage of marsupial nutritionists. 'Out of 20 PhDs I've trained, only three have carried on in the field. I continue to work closely with these three, and recently I published a new text book, *Integrative Wildlife Nutrition* (Springer 2009), with one of them. This text covers much more than marsupials, but marsupial nutrition remains my passion.'

Kangaroos, frogs, crocodiles and rockets

A zoology career has taken Professor Gordon Grigg of the University of Queensland to some unexpected places, including Mongolia, where he provided advice on adapting techniques he developed studying kangaroos to the study of endangered gazelles.

'The Mongolian gazelle is one of the last large migratory ungulates,[1] living on Mongolia's eastern steppes, one of the world's largest intact grasslands,' Grigg says. 'A combination of subsistence use and illegal poaching for "quick and dirty" sale across the borders to China and Russia are causing a decline in the gazelles.'

While Grigg has met resistance to his ideas about the sustainable harvest of kangaroos, Mongolia posed new problems. 'We were told that much of the illegal "harvest" is taken with AK-47 assault rifles and Kalashnikov submachine guns, and that the poachers are not necessarily Mongolians.'

But if this was a surprising turn of events for Grigg, it's nothing to a previous role as part of the search for the scramjet rocket lost somewhere in the Woomera prohibited area.

The Hyshot scramjet program is one of Australia's highest (in all senses of the word) technology research fields. Despite a budget many times smaller than overseas competitors receive, researchers at the University of Queensland are leading the world in producing an air-breathing propulsion system capable of operating at up to eight times the speed of sound.

In 2001, a rocket carrying a Hyshot test malfunctioned. Researchers were desperate to find the debris. Besides the potentially valuable data on board, the launch of Hyshot 2 was on hold until it was possible to determine what went wrong the first time. 'The scramjet people said the problem was the rocket; the rocket people said it was the payload,' Grigg recalls.

University of Queensland media officer Jan King had organised publicity for both Hyshot and Grigg, and thought the kangaroo aerial survey team, with Grigg as pilot, might be able to find the missing rocket. At first the rocket scientists thought this sounded like a joke. A Department of Defence helicopter

1 A category of mammals that includes horses, sheep and pigs.

had searched for many hours without luck, and they'd done their own searching in a Cessna. 'I was confident that if the rocket was in the area they calculated, our team would find it,' Grigg says.

After flying over the most likely area, Grigg found a pile of debris and guided a ground-based team to the spot. 'We saw them jumping up and down with excitement,' Grigg recalls.

The debris 'showed conclusively that the problem was with the rocket, not the payload. This laid the foundation for the successful second launch.' It was also a boost for Australian pride, since the rocket was of overseas construction.

While Grigg enjoyed this side-project 'because we succeeded,' his more regular work involves studies of animal populations. In this capacity his aerial kangaroo tracking has led him to believe that kangaroos should be seen as a resource, not a pest, and harvested for their meat and skin. The reaction from animal liberationists has been fierce, and Grigg speaks of 'entertaining brawls' in 'kangaroo politics'.

Over 25 years Grigg has built a team with the ability to count kangaroos and other large animals very accurately in a strip 200 metres wide while flying at 100 knots. He says such research, particularly over such a long time period, led to setting harvest quotas on the basis of solid information about population trends.

There is far more to Grigg's studies than aerial spotting, however. A major area of research is the impact of cane toads on native frogs in the Roper River Valley and in Kakadu through a remarkable bit of interdisciplinary teamwork.

Cane toads have a devastating effect on many Australian species when they reach a new area, but how much impact is hard to measure in the absence of baseline studies of the ecology prior to the toad's arrival. So Grigg and his colleagues have 'set up monitoring stations ahead of the toads' arrival', including voice-recognition software for frogs. When the local male frogs start calling for a mate, the solar-powered computer boots up and records the call. 'Frogs are good because each species has its own call,' Grigg explains. Once each year the researchers collect the data.

In some areas the research was not entirely successful 'because the toads arrived before we expected', providing the scientists with inadequate baseline data. The signs elsewhere are not good. 'We've had large falls in frog populations, but we are uncertain if was due to the toads. We have a second study underway at present.' It is too soon to tell if native populations will learn to adapt to the presence of the toads and slowly recover.

Grigg officially retired in 2007, but he hasn't stopped working. He's writing a book about the almost 40 years he spent researching crocodiles. 'They have possibly the most unusual heart of any vertebrate, and I collaborated on developing an understanding of how it works,' Grigg says. Crocodile and alligator hearts have an extra valve capable of directing blood away from the lungs.

Another topic that interested Grigg was the capacity of crocodiles to survive at the salty ends of estuaries during the dry season. 'I checked the [scientific] literature and it said they didn't have salt glands and I wondered how they coped. We did quite a bit of work looking at the salinity gradients in Northern Territory rivers and eventually one of my postgraduate students discovered they did have salt glands, but they're on the surface of the tongue. It was a nice puzzle to nut out.' (See page 205 for subsequent work on crocodile salt glands.)

The way crocodiles regulate their body temperature is another problem that has intrigued Grigg. 'People are very interested in the question of whether dinosaurs were warm or cold blooded, and these are the biggest living reptiles so it's interesting to look at them.' A photoessay covering Grigg's career was published in *Australian Zoologist* in 2009, emerging from a symposium held to honour his work when he retired.

Grigg was inspired to become a zoologist at age 11, 'partly because I had contact with a visiting German scientist on the Atherton Tablelands. He was a friend of my father's. I went bush with him – I think my parents were afraid he'd get lost on his own.' The visitor introduced Grigg to the wonders of pond life, after which he went on to complete a degree at the University of Queensland and postgraduate studies at the University of Oregon before returning to Australia.

There's a moth in my chocolate

Dr Terry Bowditch is the Information and Business Systems Manager at the Grains Research and Development Corporation (GRDC). However, it's his PhD that gets him into this book. For his doctorate in the University of Tasmania's Agriculture Department he became entomologist at the Cadbury Chocolate factory.

Cadbury needed an expert on insects because some were getting inside the product wrapping and consuming the precious contents. Bowditch stresses 'the rate of infestation was very low'. What is more, almost none was occurring at the factory itself.

It emerged that the longer the packs of chocolate stood at wholesale or retail outlets, the higher the rate of infestation became. The problem was predominantly in the warmer states, vindicating Cadbury's choice of Hobart as its production location.

Bowditch enjoyed the role, but not because it involved an unlimited supply of free chocolate. He says he's 'actually not that much of a chocolate person' and that there was no free stock available anyway. However, he did have access to the staff shop where product, particularly seconds, is sold at a discount.

Bowditch says he 'liked the problem-solving aspect' of his project. Every instance of infestation reported during his time was passed on to him. His first job was to identify the species involved, and then attempt to establish how it had got in. It became apparent that one species of storage moth was responsible for most occurrences, although in some cases his job was made difficult by the fact that the perpetrator had 'done the damage and left' before the box reached him.

In order to establish how these chocoholic moths were entering the packages Bowditch conducted some ecological experiments, demonstrating that the larvae were able to pick up the chocolate smell and make their way through even the tiniest holes in the packaging. Adult moths were even picking up the smell and preferentially laying their eggs near the chocolate, even if they couldn't get inside themselves.

On the management side, Bowditch suggested that the main thing was to 'ensure the packaging had good integrity' as the larvae would find even the most tiny unsealed crease or fold caused by the slightest misalignment in the machines doing the packing. There were also some cases of larvae chewing through the packaging, and he conducted studies on a more insect-proof alternative material.

Bowditch did not find a new or endangered species in any of the chocolate sent to him, but found 'the machinery involved in the manufacturing fascinating'.

The PhD topic was something of an unexpected turn for Bowditch. His undergraduate degree was in entomology at La Trobe University, but this was 'mostly evolutionary ecology' rather than pest control.

After completing his doctorate Bowditch moved on to integrated pest management (IPM) at an orchard in the Huon Valley, a location he describes as 'a beautiful place to work'. He says 'the aim was to reduce the hard [broad-spectrum] chemical usage'. Broad-spectrum pesticides work against a large number of insects, rather than targeting a particular species. It's usually cheap to attack many pests at once, but beneficial insects get killed at the same time, so the use of broad-spectrum pesticides is now discouraged as much as possible.

By the time Bowditch left after 18 months, broad-spectrum chemicals had been removed from the orchard entirely. The major pest, the light brown apple moth, was being controlled with targeted chemicals, and parasites had been introduced to the orchard to keep various species under control. 'It worked beautifully,' Bowditch says. 'The quality of fruit was much better than from orchards using chemicals.'

The GRDC undertakes no research of its own, instead assessing priorities for studies that will benefit farmers, and allocating money. Bowditch started there as a program officer before becoming program manager for the vertebrate and invertebrate pest programs.

Positions such as the one Bowditch held at Cadbury are certainly rare, but as part of his studies he discovered similar research in the US.

After leaving the GRDC Bowditch joined the Commonwealth Department of Education in Canberra, where he has worked in a number of science and research areas including the Cooperative Research Centres[1] program and the Research Block Grant programs. Bowditch is currently working on higher education policy as a Director in the Education Investment Fund branch.

1 CRCs bring together universities, institutions such as CSIRO, government and industry to make scientific research relevant to the needs of industry.

Birdcatching on the fly

Dr Clive Minton says he's been 'interested in outdoor things, particularly birds, ever since I was hatched'. Nevertheless, the pioneer of studies of wading birds in Australia did not make a career from ornithology.

With a PhD in metallurgy from Cambridge University, Minton helped develop titanium alloys that are still used in some aircraft today. His employer, IMI Metals, eventually sent him to Australia, but after five years requested he return to England. 'I told them I'd torn up my return ticket,' Minton says. For an ornithologist, Australia was just too tempting a place to live.

Minton moved away from scientific employment, becoming in turn director of human resources at Myer, Deputy Head of the Health Department of Victoria, and leader of a head-hunting agency, before retiring to write up the data he'd collected on migratory wading birds since he arrived in Australia.

While still in England, Minton had taken a young sanderling he could not identify to Dr Eric Ennion, an expert in wading birds. Ennion inspired Minton's studies of wading birds, through which Minton met his wife. To capture the birds Minton adopted and eventually improved the technique of 'cannon netting', in which small explosives are detonated behind a net the size of a tennis court, carrying it over some rather surprised wading birds.

This technique enabled ornithologists at The Wash, the largest area for wading birds in the British Isles, to increase the number of birds they caught and tagged (banded) from a few hundred to several thousand per year. 'Having these established techniques and experience, coming to a greenfield such as Australia, where almost nothing was known about wading birds, was like entering the Garden of Eden,' Minton says.

Those interested in joining the thousands participating in the world-leading banding project should contact Birds Australia, but Minton warns long days are required, sometimes in inclement conditions. Team members must swiftly lift the front of the net (birds and all) out of the water and bring the birds to land.

The birds are stored in holding compartments under shade cloth until they can be weighed, banded and released. 'We have to make sure we keep to a minimum any negative effects [on the birds],' Dr Minton says. 'But you can't do science on wildlife without some risk of damage.'

Of Australia's 73 species of wading birds 55 are migrants, many breeding in the Arctic Circle before flying 12 000 km to spend the majority of their time in Australia. The average life expectancy of these birds is just 3–5 years, but some individuals live to 20. By that time, Dr Minton notes, birds 'weighing as little as 30 grams and fitting in a wine glass' will have flown 500 000 km, 'more than the distance to the Moon'.

The Arctic summer is so short that the birds have little time between hatching and flying south. 'In some species the parents leave their offspring before they can fly, only keeping them warm for a few days and teaching them to avoid predators,' Minton says. 'In the egg there is genetic code saying "fly south-south-east, my son, for 12 000 km and there you will find Melbourne".'

Minton has authored more that 100 scientific papers and articles on wading birds, and was made a Member in the Order of Australia (AM) for his services to ornithology. Not surprisingly, he is passionate about the threat posed by damage to the birds' East Asian stopping points. 'Knowledge is the foundation for conservation,' he says. The research he and others have done has prompted agreements with China and Japan to protect some wetlands in which the birds 'refuel' for their amazing journeys.

Minton was involved in a campaign to prevent the draining of a huge estuary in South Korea. While that campaign was unsuccessful, Minton hopes 'The furore which has been created and information about the damage done will prevent [the government] doing similar things elsewhere.

'Major reclamations of inter-tidal areas have continued on the Korean and Chinese shores of the Yellow Sea and it is now having a major impact on the numbers of many species of migratory waders here in Australia,' Minton says. 'We have recorded major reductions in the populations all over Australia during the last three years and the decline is ongoing. We are still hoping our data can be used more effectively by governments and international conservation agencies to try and ensure the protection of the remaining inter-tidal areas particularly the "hot spots" used by these birds.'

Minton was fortunate in having an interest in an area of science in which it was possible for amateurs to make a major contribution. 'At 13–14 I had to decide whether to become a biologist or follow [other areas of science]. Many of my friends have done well out of going into birds, but I think that I have had even greater enjoyment because my hobby complemented my career. The fact that they were so different made for a delightful contrast.'

Crocs and bum-breathing turtles

Some academics have expressed concern that student research projects are becoming too safe, but they're probably not thinking of the work of ecophysiology Honours student Inga de Vries. Sticking your hand into the mouth of estuarine crocodiles (*Crocodylus porosus*) may be described as many things, but excessively safe doesn't spring to mind.

Admittedly, de Vries' subjects are juveniles under 1 metre long but 'you still wouldn't want them to latch on to you', as she puts it. The crocs in question were housed in an animal centre and restrained, but, de Vries notes, 'They were conscious at the time, not drugged.'

The motivation for de Vries' actions was to investigate the salt glands on the crocodiles' tongues. It is these glands that enable saltwater crocodiles to spend long periods of time in the ocean. They extract salt from their blood and concentrate it into a salty solution that is then excreted through pores onto the surface of the tongue, where it can be washed away. Little is known about how these glands work; their discovery is recent (p. 198) and de Vries was endeavouring to find out.

The glands get bigger and more active if the crocodiles spend long periods of time in marine environments, and seem to become inactive when they are in fresh water. Similar glands exist in birds such as albatross, enabling them to drink seawater while on long journeys. De Vries is not sure whether these glands were inherited from a common ancestor or are an example of 'convergent evolution' whereby different species independently evolve the same or similar biological traits in response to the same pressures. Australian freshwater crocs also have the glands, enabling them to spend time in salty areas when necessary, although they generally prefer not to.

An even more interesting evolutionary question was posed by de Vries' previous research topic – freshwater turtles that can absorb oxygen from water through their cloaca. As de Vries puts it: 'They can breathe through their butt.'

Like other reptiles (and most birds), the turtles' cloaca is the only opening for their intestinal, urinary and genital tract. However, the turtles' cloacas have added an extra function, being capable of taking water into thin-walled sacs from which dissolved oxygen can be removed.

No known turtle can survive using cloacal respiration for all its oxygen, but de Vries says that some species can get 4% of their oxygen requirement from the sacs while others can get up to 70%. De Vries studied the effects of water temperature and depth on turtle behaviour. She says, 'It's really interesting because there are lots of trade-offs. It uses a lot of energy to breathe cloacally, but so does surfacing. Also, in the wild the turtle is probably safer resting on the bottom than at the surface.

'When it was cold they usually hung around the bottom,' de Vries observes, 'but as it warmed up they moved around more.' When the water was deeper the turtles were more inclined to stay on the bottom – presumably deciding that the surface was too far away.

The turtles have achieved some fame because a species that lives on Queensland's Mary River was threatened by a proposal to build a dam across the river. Deep, still water has less air dissolved in it than water that's been through rapids, which means a dam would make it harder for the turtles to draw the oxygen they need. De Vries' research carries implications for the ability of the turtles to survive changes to river flow. The dam proposal was rejected by the Federal government. The Queensland government considered appealing, but at the time of writing have not done so.

The remarkable thing about de Vries' turtle experiments is that they were done while she was an undergraduate student. The University of Queensland has programs for those who are keen on science and doing well. This allows them to get into the labs and conduct real ground-breaking research rather than repeating experiments done by thousands before them, as usually happens in undergraduate laboratories. (Alecia Carter is another beneficiary of this approach; p. 178.) De Vries is keen to credit Professor Craig Franklin of the School of Biological Sciences who supervised her work on both the turtles and crocodiles.

De Vries' first experience in this regard was conducting side research on the pattern of tail markings of a species of cleaner fish. She established that the colouring followed a geographic distribution along the Queensland coastline. After this she conducted a study of the lifecycle of a parasite.

Animal research was always a natural path for de Vries, who says: 'At least since the start of high school I wanted to be a veterinary scientist.' However, a career as a vet was somewhat hampered when she became allergic to many mammals. Consequently she decided to go into marine research, thinking 'I could still hang out with animals.'

De Vries found a course in animal physiology fascinating, and as a result switched her research focus, first to the turtles and then to the crocodiles located in the same lab.

After finishing her Honours de Vries began working as an environmental officer for Queensland Rail. 'I've been working there the last two and a half years,' she says. 'While this job is unrelated to lab work, I use many of the skills I learnt while working at uni. I do a lot of data collation, data analysis, project management and report writing, among other things.'

'My field of interest has changed as well. While at uni I was focused on physiology of animals, I am now interested in environmental management. Currently I have taken time off work for three months to travel through South America (including volunteering in an animal refuge – I still love animals!). I plan to return to my Environmental Officer position and would like to learn more about environmental management. I am also interested in getting involved in teaching/training, and just before I left I started a course to become a professional trainer. I plan to use this in my job to train other staff on their environmental responsibilities.'

www.ingramcontent.com/pod-product-compliance
Lightning Source LLC
LaVergne TN
LVHW050532100826
845148LV00002B/526